Jane's
Facility
Security
Handbook

Christopher Kozlow

John Sullivan

Published by Jane's Information Group
1340 Braddock Place, Suite 300 Alexandria, Virginia
22314-1651, USA
Tel: 703-683-3700; Fax: 703-836-0029; E-Mail: info@
janes. com

Printed in the USA

**Registered with the Library of Congress
Cataloging-in-Publication Data available upon request**

ISBN 0-7106-2288-0

Publisher's Note
Jane's would like to thank the two authors of Jane's Facility Security Handbook, Christopher Kozlow and John Sullivan, for their tireless effort. Christopher Kozlow authored the Hospitals, Educational Institutions, Utilities and Entertainment Facilities chapters of the book. John Sullivan was responsible for the Transportation Systems and Special Events chapters. Applying their years of experience in the fields of first response, emergency management and physical security, these gentlemen produced a first-rate manual.

This volume also contains information gleaned from the following Jane's Publications:

♦ Jane's Chem-Bio Handbook
♦ Jane's Chemical and Biological Defense Guidebook, Second Edition
♦ Jane's Counter Terrorism
♦ Jane's World Insurgency and Terrorism

Brian Marshall
Coordinating Editor
Terrorism, Insurgency and Intelligence
Jane's Information Group

TABLE OF CONTENTS

Chapter I: Terrorism Primer

Terrorism: Definitions and Terms

The Terrorist Threat 3
Terrorism Defined 5
Counter Terrorism Defined 9
Terrorist Organization and Structure 12
Typical Terrorist Organization 14
Types of Terrorist Groups 16
Terrorist Objectives 17

Terrorist Tactics and Targets

Terrorist Tactics 20
Potential Terrorist Targets 30

Chapter II: Security Planning

Facility Security Planning

Introduction 39
Performing a Security Survey 40
Planning for Bomb Threats and Incidents 44
Search Teams 50
Anti-Terrorism Survey 55
Workplace Security Survey 58

Chapter III: Buildings Security

Threats to Buildings from Terrorist Attack

Introduction 73
Likely Target Structures 75
Terrorist Threats to Building Security 77
Why Buildings are Vulnerable to Terrorist Attack 83
Vulnerability Assessment: Building Profile 91

Vulnerability Assessment Tools and Techniques 94
Hardening a Target 97
Securing Access and Entry Points 101

Chapter IV: Hospitals

Hospital Security Planning 107
Step One: The Planning Process 108
Step Two: Planning Checklists 111
Step Three: Response Folders 114
Step Four: Emergency Tasks 122
Event Recognition 126
Hospital Incident Management 127
Public Safety Response 132
Planning for Different Types of Attacks 137

Chapter V: Educational Institutions

Educational Institutions
Definitions 141
It Could Happen! What Do We Do First? 144
Vulnerability and Hazard Analysis 146
Action Planning Checklist 149
Responding to Crisis 151
Direction and Control 155
Command Post 157
Overall School Responsibility 161
Before Help Arrives 165
Emergency Procedures
Direction and Control 166
Responding in the Aftermath of Crisis 170
Crisis Procedure Checklist 172

Chapter VI: Transportation Systems

Introduction 175
Transportation Infrastructure 175
Security Issues and Transport Systems 177
Response to a Crisis, Attack or Disaster 179
Initial Assessment 181
Target Folder/Response Information Folder 184
Terrorism and Attacks Against Transit Systems 185
Recording Threats 189
Searches 191
Incident Objectives for Terrorist Attack 193
Scene Security Concerns 194
CBRN Incident Indicators 198

Chapter VII: Utilities

Introduction 203
Utility Companies as Targets 203
Step One: The Planning Process 206
Step Two: Planning Checklists 207
Step Three: Response Folders 212
Step Four: Emergency Tasks 218
Terrorist Weapon Systems 222

Chapter VIII: Entertainment Facilities

Introduction 231
Planning Steps 231
Step One: The Planning Process 232
Step Two: Planning Checklists 233
Step Three: Response Folders 240
Step Four: Initial Actions On-Scene 246
Immediate Actions 248
Bomb Threats and Bomb Incidents 251
Possible Weapon Types and Tactics 253

Chapter IX: Special Events

Introduction 259
Terrorism and Special Events 260
Public Order and Special Events 261
Sports Violence and Special Events 262
Planning for Special Events 263
Threat Assessments 268
Minimum Intelligence Requirements 272
Disorder and Special Events 275
Crowd Composition 277
Chemical, Biological and Radiological Attacks 281
CBRN Incident Indicators 283

Chapter X: Response

Response to a Terrorist Incident
First Response Incident Phases
Introduction 287
First Response Incident Phases 288
Initial and Response Action 292

Management of a Terrorist Incident 302
History of Incident Management and Command 304
Incident Commander Role and Responsibilities 309

Appendix A: Glossary of Acronyms 317

Chapter I: Terrorism Primer

Terrorism: Definitions and Terms

The Terrorist Threat

Terrorism Defined

Counter Terrorism Defined

Terrorist Organization and Structure

Typical Terrorist Organization

Types of Terrorist Groups

Terrorist Objectives

Terrorist Tactics and Targets

Terrorist Tactics

Potential Terrorist Targets

Terrorism: Definitions and Terms

THE TERRORIST THREAT

International Terrorism

At a macro-level, the trend in recent years has been towards an overall decrease in the extent of international terrorism. The number of international incidents fell from a high of 665 in 1987 to only 304 incidents in 1997. This marked an increase by eight incidents over 1996, but the numbers nevertheless reflect a general downward trend.

However, the numbers belie the seriousness of the problem **political violence** poses today. Deaths resulting from **international** and **domestic** terrorism and insurgency are on the increase. Precise evidence is elusive and often inaccurate, however, largely because there are no internationally accepted definitions for terrorism, insurgency and political violence.

Domestic Terrorism in the US

In the recent past, the US population at large had the opinion that terrorism and terrorist attacks could never take place on US soil. In fact terrorism has been occurring in the US for decades. Only recently has the US population and the **emergency response community** at large started to realize that the spectacular attacks which have occurred in the US resemble tactics associated with international groups overseas.

The bombing of New York City's **World Trade Center** on 26 February 1993 made it clear that the US is no longer invulnerable to terrorism. Terrorism can and inevitably will occur in the US. The bombing of the **Alfred P. Murrah federal building** in Oklahoma City on 19 April 1995 subsequently changed most Americans' attitudes regarding the threat posed by domestic terrorism.

US law enforcement authorities and security planners need to be prepared to respond to both **international** and **domestic terrorist incidents**. The number of extremists, anti-government movements, left– and right-wing terrorists and other special interest groups in the US rose dramatically in the 1990s.

The attitudes and potential actions of such international and domestic groups are of concern to law enforcement and intelligence specialists, since extremists, anti-government groups and other supporters of terrorism are forming world-wide alliances and information-sharing networks to present a changing threat environment. This has given law enforcement officials considerable cause for concern.

TERRORISM DEFINED

According to the US Department of State, there is no one universally accepted definition of terrorism. Different definitions hold sway in academic, military, law enforcement and legal settings. For the purposes of this work and emergency planners and responders who might have to deal with terrorism, the US Federal Bureau of Investigation (FBI) uses the following definition.

Terrorism involves:

♦ The unlawful use of force or violence against persons or property to intimidate or coerce a government, the civilian population, or any segment thereof, in furtherance of political or social objectives.

There are other definitions of terrorism, however, and there are differences between definitions of terrorism even within the same government.

The US Department of State (DoS) provides the following definitions of terrorism under Title 22 of the United States Code, Section 265(f) d:

♦ The term **terrorism** means premeditated, politically motivated violence perpetrated against non-combatant targets by sub-national groups or clandestine agents, usually intended to influence an audience.

♦ The term **international terrorism** means terrorism involving citizens or the territory of more than one country.

♦ The term **terrorist group** means any group practicing, or that has significant subgroups that practice, international terrorism.

The US Department of Defense (DoD) defines terrorism as:

♦ The unlawful use of — or threatened use of — force or violence against individuals or property to coerce or intimidate governments or societies. Often used to achieve political, religious or ideological objectives.

Undoubtedly the word terrorism evokes strong emotions, and it is a politically sensitive term in all countries. Clearly no single definition of terrorism can be satisfactory to all. However, most acts of terrorism betray certain constants all around the world. Terrorism can and often does strike civilians — not merely government employees or political activists. In short, **terrorism is a technique for inducing fear by intimidation**. Terrorists generally act very deliberately, and as such acts of terrorist violence are neither spontaneous nor random. Careful planning and execution is the hallmark of most terrorist attacks.

The FBI distinguishes between three distinct categories of terrorist-related activity:

♦ A **terrorist incident** is a violent act or acts dangerous to human life, in violation of the criminal laws of the United States or of any state, to intimidate or coerce a government, the civilian population or any segment thereof, in furtherance of political or social objectives.
♦ A **suspected terrorist incident** is a potential act of terrorism in which responsibility for the act cannot be attributed at the time to a known or suspected terrorist group or individual.
♦ A **terrorist prevention** is a documented instance in which a violent act by a known or suspected terrorist group or individual with the means and a proven propensity for violence is successfully interdicted through investigative activity.

Domestic Terrorism Defined
Domestic terrorism in the US context involves groups or individuals based and operating entirely within the United States, Puerto Rico, and other US territories, without foreign direction, whose acts are directed at elements of the US government or the US population.

International Terrorism Defined
International terrorism is the unlawful use of force or violence by a group or individual connected to a foreign power, or whose activities transcend national

boundaries, against persons or property to intimidate or coerce a government or the civilian population in furtherance of political or social objectives.

COUNTER TERRORISM DEFINED

Anti-terrorism is defined as:

♦ Defensive measures used to reduce the vulnerability of individuals and property to terrorist acts, to include limited response and containment by local military forces.

Counter terrorism is defined as:

♦ Offensive measures taken to prevent, deter and respond to terrorism.

US Presidential Decision Directive 39 (PDD-39), published in June 1995, established policy for US counter-terrorism efforts. PDD-39 states:

♦ Terrorism is both a threat to our national security as well as a criminal act. The administration has stated that it is the policy of the United States to use all appropriate means to deter, defeat and respond to all terrorist attacks on our territory and resources, both people and facilities, wherever they occur.

In support of these efforts, the US will:

♦ Employ efforts to deter, pre-empt, apprehend and prosecute terrorists.

- ◆ Work closely with other governments to carry out US counter-terrorism policy and combat terrorist threats.
- ◆ Identify sponsors of terrorism, isolate them and ensure they pay for their actions.
- ◆ Make no concessions to terrorists.

According to PDD-39, the US will (1) make no deals with terrorist groups or nations supporting terrorism; (2) treat terrorists as criminals, pursue them aggressively and apply the rule of law. (3) Terrorist acts are criminal, whether in peacetime or war. In peacetime, the government will prosecute terrorists for violating the criminal laws of the state and country in which they commit their crime.

The US will also apply maximum pressure on states that sponsor and support terrorists through economic, diplomatic and political sanctions, will urge other states to follow suit. PDD-39 designated the FBI the **Lead Federal Agency (LFA)** in crisis management response to terrorist activity.

The FBI has responsibility for law enforcement regarding terrorist groups in the US and acts of terrorism directed at Americans overseas.

However, this role is chiefly as a coordinator. In most instances the FBI only assumes a lead position. The agency employs subject experts to manage technical matters beyond FBI operational abilities, and operates in conjunction with local first responders.

Table I: US Federal Counter-Terrorist Activities		
Prevention and Planning	**Response**	
Identify and monitor potential threats	Communicate with the media and public	Provide shelter and services
Raise public awareness	Suppress/counter immediate threat	Maintain continuity of federal government
Assess and enhance readiness	Evacuate citizens in the threat area	Initiate criminal investigation
Prepare written contingency plans	Rescue lives and save property	Assess damage
Secure potential targets and threat interdiction	Provide emergency medical assistance	Undertake remediation and restoration
General support efforts	Maintain/restore domestic order	Provide compensation and assist victims

Chapter I: Terrorism Primer

TERRORIST ORGANIZATION AND STRUCTURE

Many modern terrorist organizations resemble modern corporations, with established rules and regulations and worldwide financial investors, much like stockholders in large corporations.

Some terrorist organizations establish clear **lines of authority** and **command**. Terrorist groups establish these organizational structures in an effort to provide support, training, decision making and other resources to active group members operating in frequently hostile environments.

Yet in most terrorist organizations, group dynamics, philosophical and ideological differences can override the organizational structure. Individual members, or **factions**, may carry out attacks that contravene the directives of the group's central authority. For example, some members of the US **Patriot movement** believed the Oklahoma City federal building bombing was excessive, and the associated deaths needless. Such internal conflicts can cause smaller factions to splinter off and form new terrorist groups, which while distinct entities, still maintain the support of the larger terrorist organization.

Terrorist groups, especially those without considerable government support, generally need clear structures. Typical terrorist groups include a core group of operational members, comprised of the leadership and active terrorists, as well as additional supporters and sympathizers.

Cellular Structure

With increased anti-/counter-terrorism initiatives worldwide, new forms of terrorist organizations and structures are increasingly smaller or **cellular** in nature. Individual cells are relatively isolated from other cells within the organization. This eliminates the extent to which authorities can damage the larger organization if they infiltrate a single cell or capture cell members. Defections or captures yield limited benefits. **Compartmentalization** ensures that no terrorist can identify more than a few others.

The **multi-functional cell** is a new form of cell structure visible in the US anti-government and extremist movements. These cells consist of skilled individuals, trained in one or more different skills in guerrilla warfare. These small tactical cells can operate independently of the organized command structure, increasing the likelihood of violent acts without prior warning or detection.

TYPICAL TERRORIST ORGANIZATION

◆ Leadership
◆ Active Terrorists
◆ Active Supporters
◆ Passive Supporters

Leadership
The leadership within a terrorist organization is generally at the top of the command and control structure. Its primary role in the organization is to define policy, set objectives and to direct actions. The leadership is intensely committed to the group's causes and often consists of charismatic individuals who have the ability to motivate others. If a group is **state-supported** or **-directed**, the sponsoring state frequently trains and educates some or all of the leadership.

Active Terrorists
Active terrorists are usually recruited and trained by the organization in methods of carrying out attacks. Many active terrorists are deeply committed to the group's cause, while others within the general membership may be 'professional' terrorists, who may not be motivated by ideology.

Active Supporters
Active supporters usually do not actually commit attacks for the organization. They do provide money, intelligence, safe houses, forged or stolen

documents, legal assistance and/or medical care. Active supporters tend to agree with some or all of the group's goals and causes; they may be ambivalent regarding the use of violence. Most organizations recruit active members from within this group, since active supporters prove their loyalty and skills to the organization and leadership. A certain number of active supporters are thrill seekers who join predominantly for the excitement of belonging to a forbidden organization.

Passive Supporters

Passive supporters are more difficult to define and identify. Most are sympathetic to the terrorist cause, but may not assume an active role within the organization. Some passive supporters become involved in intimidation and extortion operations. Most terrorist organizations depend on passive support for financial backing and logistical operations.

TYPES OF TERRORIST GROUPS

Non-State Sponsored Groups
Non-state sponsored terrorist groups operate autonomously, and receive no significant support from any government. Italy's **Red Brigades** and Spain's **Basque Fatherland and Liberty (ETA)** are examples of non-state supported groups.

State-Supported Groups
State-supported terrorist groups generally operate independently, but still receive government support. The Lebanese **Party of God (Hizbullah)** is an example of a state-supported group.

State-Directed Groups
These groups or organizations operate as direct government agents. Such terrorist groups receive intelligence and operational support from the directing government. Libyan 'hit teams' that once targeted Libyan exiles are classic examples of this form of terrorism.

TERRORIST OBJECTIVES

Analysts can classify terrorist attacks by immediate terrorist objectives. Common objectives can include recognition, coercion, intimidation, provocation and insurgency support. It is not uncommon for a terrorist group to pursue one or all of these objectives simultaneously.

Recognition

From the beginning of a terrorist campaign, the objective of an attack or series of attacks may be to gain domestic or international recognition for a cause. Recognition can bring **support** and **funds** from outside sources, and aid efforts to **recruit** new members. An attack can also demonstrate strength, proclaim a group's existence and demonstrate its ability to attack anywhere.

Groups attempting to gain recognition generally execute spectacular attacks that attract significant media attention. They may also adopt names or labels that imply strength or mass of numbers (such as Front, Army or Brigade).

Coercion

Coercion is an attempt to induce a desired behavior or action from individuals, groups or governments. This objective calls for a strategy of selective targeting that can cause massive destruction, but usually not large numbers of casualties. Examples can include bombings of a corporate headquarters of a company involved in environmental operations, or the bombing of abortion clinics.

Intimidation

While coercion is an attempt to force an action, intimidation is designed to prevent individuals or groups from acting. A strategy of intimidation aims to so frighten security forces that their effectiveness is reduced. The mere threat of terrorist retaliation can foment fear and cause individuals to alter their daily routines.

Provocation

Provocation is an endeavor to trigger a heavy-handed response from a government agency or military. This strategy can involve terrorist attacks against symbols of a government, to demonstrate a particular government's vulnerability. The objective is two-fold. Terrorists hope provocation attacks will decrease public faith in the government's ability to shield the populace from threats. Terrorists may also seek to win compassion and public sympathy, if the government launches a counter offensive.

Insurgency Support

Terrorists can support insurgencies' quest for recognition with acts of provocation, intimidation and coercion. Focused terrorist strikes can lead a besieged government to overextend itself, and try to protect all potential targets. Insurgency support can also include fundraising, coercion of recruits, logistical support and enforcement of internal group discipline.

Terrorist Tactics and Targets

Terrorist Tactics

At a fundamental level, all terrorists seek to create a sense of fear to further their cause or achieve political or other objectives. Some groups focus on one or two different tactics, while others show greater variety. The tactic that a group employs depends not only on its goals and operational capabilities, but also on the terrorist group's financial limitations.

Traditional terrorist tactics include:

♦ Assassination.
♦ Armed assault.
♦ Bombing.
♦ Hijacking.
♦ Hostage-taking.
♦ Kidnapping.

Newer terrorist tactics include:

♦ Product contamination.
♦ Cyber-terrorism.

Assassination
♦ Assassination is the planned murder of an individual, whether a senior government official, first responder or military member.

Table II: Basic and Advanced Training Skills	
Basic Training Skills	**Advanced Training Skills**
Weapons handling — small and large	Intelligence gathering
Explosives production and construction	Encryption and decryption skills
Land navigation	Demolition facilities
Survival skills	Flight training
Secure communications	Driving skills
Escape and evasion	Computer training
Ambush skills	Weapons of mass destruction production
Mountaineering	
Clandestine travel techniques including:- False document production - Weapons and explosive smuggling - Disguises	

Observers often perceive political murder and assassination to be interchangeable terms. The commonly accepted definition of assassination has typically been the murder of a head of state or other political official.

However, assassination is no longer an act of terrorism solely or even primarily directed at heads of state. Recent assassination targets include military or

emergency response personnel, seen as legitimate targets and not innocent victims.

Assassination itself can take many different forms, including stabbing, shooting, poisoning or bombing. Terrorists can use stand-off weapons to carry out assassinations from a distance.

These weapons can include:

◆ Remote explosives detonation.
◆ High-powered rifles.
◆ Missiles.
◆ Rockets.
◆ Mortars.

Assassinations are rarely haphazard events. Terrorists carefully select their targets for specific strategic purposes. Assassinations also generally require some degree of organizational structure, intelligence collection capability, weapons support and financial backing.

Terrorist assassins usually receive training in:

◆ Security system penetration.
◆ Clandestine approach.
◆ Disguises.
◆ Escape and evasion techniques.

Assassins are typically committed to their cause and some may be willing to die in order to complete their missions.

Armed Assaults

♦ Armed assault is a terrorist tactic that employs surprise, speed and heavy weaponry.

Armed assault is a tactic terrorists use to strike selected targets with little warning, limited planning and resources. One method terrorists employ when carrying out armed assaults is the stand-off attack.

Common stand-off weapons used in armed attacks include:

♦ Mortars.
♦ Rockets.
♦ Remotely detonated explosive devices.

This distant form of attack allows terrorist groups to strike targets from outside the operational control of security forces. The **Provisional Irish Republican Army (PIRA)** successfully employed this tactic in the past, carrying out effective attacks on selected targets with a minimal risk of capture.

Another form of armed assault is the close-range strike, executed very rapidly and with deadly effectiveness. The **Abu Nidal Organisation (ANO)** carried out attacks on the Rome and Vienna airports in December 1985.

Bombings

♦ Bomb attacks employ explosive devices, rigged to explode on impact or on a timed delay, to make a political statement.

Bombs have long been terrorist weapons, at least as far back as anarchist attacks in Russia and the US during the 1880s. Several factors make bombs popular weapons:

1. Terrorists can fairly easily acquire explosives and associated technology at a relatively low financial cost, while device manufacture and construction require little technical expertise.
2. Readily available "how-to manuals" make the task even simpler.
3. Bombings are clandestine and low-risk attacks. Sophisticated timing devices enable terrorists to leave the scene before devices can explode.
4. Bombing as a terrorist tactic is unparalleled as a media event, and the larger the explosion, the larger is the level of media coverage.

Hijacking

♦ Hijacking is frequently a tactic designed to gain headlines, limited operationally by the size and financial resources of a terrorist individual or group.

Aircraft hijackings offer terrorists worldwide media attention, mobility and usually reflect highly-focused objectives. Terrorists also sometimes hijack trains, buses and seagoing craft. In October 1985, the Abu Abbas faction of the **Palestine Liberation Front (PLF)** hijacked the Italian pleasure cruise ship *Achille Lauro* to secure the release of Abu Abbas from Italian custody.

Hostage-Taking

♦ Hostage-taking entails the overt seizure of one or more persons to gain publicity, concessions, ransom or the release of imprisoned comrades.

Terrorists usually plan hostage situations very carefully and employ extensive prior surveillance and reconnaissance. Typically, domestic groups only seize hostages in the local environment.

In search of publicity, terrorists may target individuals whose detention will attract attention due to their political status or fame. In these instances, the event may play to a national and/or international audience.

Table III: Kidnapping Motives						
	Motive					
Target	Publicity	Political Concession	Prisoner Release	Raise Funds	Make Example	Force Act
Business				X		
Wealthy				X		
Military		X	X			
Government	X	X	X			
Opponents	X				X	
Other				X		X

Kidnapping

Kidnapping involves the seizure of prominent government officials, businessmen or other targets. Kidnappers hold their hostages until a government or corporation meets a specific set of demands.

Although similar, there are several significant differences between kidnapping and hostage-taking. Kidnapping requires elaborate planning and logistics, and is usually covert. The perpetrators of a kidnapping also may not make themselves known to authorities for an extended period of time.

1. Media attention associated with kidnapping is often less intense than in hostage instances, since there is a far more extended time factor.

2. Hostage-taking usually presents a far lower risk of capture than do other terrorist tactics.
3. An increase in kidnappings in a region often precedes larger terrorist operations:

Kidnap and ransom operations frequently finance future terrorist or insurgency operations. Table III provides a list of common kidnapping targets, and the associated goals of kidnapping.

Product Tampering

♦ Product tampering involves terrorist alterations in the elements of commercial products or services, such as the injection of poison in food or pharmaceutical products manufacturing.

Terrorists can employ product tampering:

♦ As an extortion method for monetary gain; and
♦ To undermine faith in a particular government.

Even the threat of product tampering alone can occasionally induce a government or business to remove products from distribution to the public, resulting in the loss of revenue and confidence.

Other potential terrorist tactics include the following:

Arson presents terrorists with a low risk act, and requires a low level of technology. Terrorists can also easily disclaim arson attacks.

Sabotage can demonstrate the vulnerability of a society to terrorist actions.

♦ Utilities (including electricity, gas and water), communications and transportation systems are so interdependent that a serious disruption in one can gain public attention instantaneously.

Hoaxes and threats also gain public attention.

■ Bomb threats can close commercial buildings, empty theaters and disrupt transportation systems, at no cost.

■ Repeated threats can also dull the operational effectiveness of emergency response personnel, which can in turn create opportunities for legitimate terrorist strikes.

Weapons of Mass Destruction (WMD)
One definition for Weapons of Mass Destruction (WMD) is the following:

♦ Weapons that include nuclear explosives or radiological contaminants, lethal chemicals or lethal
  biological agents (toxins or pathogens).

Few terrorist threats generated the degree of near
panic around the world in the late 1990s, as did concerns about the use of WMD as terrorist weapons.
The **Tokyo sarin gas attack** of 1995 is only the most
serious incident of **chemical terrorism** on record.
There have been numerous recorded instances
when authorities have foiled terrorist attempts to
acquire or to use **chemical-biological (CB) weapons**, while anthrax **hoaxes**, or threats, have become
relatively commonplace in the US in particular.

Well documented lapses in Russian security controls over the former Soviet nuclear stockpile, and
numerous arrests of nuclear smugglers in Eastern
Europe, have also generated fears that terrorists may
be able to procure nuclear material to construct
radiological weapons for use against the West.

POTENTIAL TERRORIST TARGETS

It is virtually impossible to develop an all-inclusive list of potential terrorism targets. In today's urban environment target selection is nearly unlimited, though in most cases potential targets have obvious symbolic traits. Terrorist targets may:

♦ Represent a government or institution, such as a federal building.
♦ Be the symbol of a large city.
♦ Be a politically sensitive business, such as an abortion clinic or a bar frequented by homosexuals.

Regardless of the target, in nearly all cases terrorists are very selective in target selection. (Table IV.)

■ Many, though not all, terrorists select targets that offer a minimal risk of capture.

■ Attractive targets are easily accessible; terrorists can easily overcome the physical and personnel security measures that are in place.

Successful terrorists learn from past mistakes. Increasingly, terrorist groups communicate with each another to exchange information regarding target selection. This exchange about previously planned or attempted attacks can be relevant for future strikes. This was the case in the US Oklahoma City bombing.

Table IV: Potential Terrorist Targets		
Persons	**Places**	**Things**
Elected government officials	Airports	Computer terminals
Federal, state and local officials	Federal, state and local facilities	Utilities, pipelines, power
Military officials and personnel	Sporting events	Cruise ships
Public safety officials	Military bases	Satellite transmitters
Business leaders	Universities	Vehicles
Government workers	Planned events (parades)	Radio antennas

(For further information see **Chapter III,** Building Security)

Certain *specific* terrorist targets can also cause law enforcement professionals considerable concern:

Hospitals
Of especial concern to emergency responders are terrorist attacks against hospitals. Following an initial bomb or chemical-biological (CB) weapon attack, hospitals may become overwhelmed with victims and first response personnel.

■ An explosive, or agent release, at a hospital after an earlier attack can raise the death count.

■ Such attacks enable terrorists to target first responders and survivors of earlier attacks.

(For further information see **Chapter IV,** Hospitals)

Educational Institutions

Schools and universities have been the scene of numerous shootings in recent years in the US, by students who have brought firearms onto campus.

Terrorists also understand that young students make attractive hostages, and can help terrorists gain media attention. The ease with which terrorists can bring weapons onto campuses is another advantage.

■ Many university and secondary institutions cover expansive and open campuses, which pose unique security problems for security personnel.

The Lord's Resistance Army (LRA) regularly attacks secondary schools in Uganda, and kidnaps students for ransom and forced recruits.

(For further information see **Chapter V,** Educational Institutions)

Airline Industry

The airline industry has long been a favored terrorist target, since it:

♦ Exemplifies major economic activity in a number of countries around the world.
♦ Is easily disrupted and has high operating costs.
♦ Offers a wide array of hostages, of all nationalities, whose seizure is sure to attract worldwide media attention.

The ultimate goal of this economically driven form of terrorism is to upset travel and tourist activity.

(For further information see **Chapter VI,** Transportation Systems)

Utilities

For some time, utility companies have been concerned about the possibility that terrorist attacks might affect their operations. This concern is well-warranted. Electrical power, communications, water systems as well as gas and oil pipelines are all vulnerable and present attractive targets.

(For further information see **Chapter VII,** Utilities)

Electricity Utilities

Electric power grids are highly vulnerable and thus have become top targets for terrorist organizations.

Terrorists understand that power grids are built-in loops. While this feature allows grids to withstand multiple power outages and failures, it is also an advantage to terrorists.

■ Terrorists can identify 'choke points' in the power grid and attack designated vulnerable spots to cause wide-scale blackouts over large areas of the country. They can then plan out attacks in affected regions where response personnel are dealing with blackouts.

Sendero Luminoso (SL), the Shining Path, launched highly effective terrorist strikes in Perú against Lima's power grid in the 1980s and 1990s.

Gas and Oil Utilities

Oil and gas pipeline systems are also vulnerable networks that are very difficult to secure — a fact that terrorists understand all too well. Energy resource extraction and distribution systems are vulnerable both at sea and on land.

Terrorists can conduct clandestine operations against offshore platforms, using:

1. Suicide bombers who ride pleasure craft;
2. Scuba divers who place explosives or mines on a platform base; or
3. Kamikaze-style aircraft.

Chapter I: Terrorism Primer   www.janes.com

Once gas or oil reaches land via pipeline networks, it becomes highly susceptible to attack. Established safety precautions require pipeline operators to label and identify their pipes and product every time pipelines cross a road or waterway. These safety procedures enable terrorists to detonate sections of pipeline that carry volatile substances.

In Colombia, the Ejército de Liberación Nacional (ELN), the National Liberation Army, continues to inflict significant economic damage and gains significant exposure from its regular bombing campaign against the Cano Limon-Covenas oil and natural gas pipeline.

Entertainment Facilities

First response communities have started to examine the potential for terrorist attacks at large sporting events, concerts and activities in stadiums, arenas and concert halls.

Chemical or biological weapons attacks at sports events are far from fictional scenarios. Indoor arenas can hold from 15,000 to 27,000 spectators, and capacity at stadiums is far higher. Such masses of people in close proximity provide the perfect opportunity for terrorists to cause a **Mass Casualty Incident (MCI)**.

■ Prior to sports events or concerts, terrorists can place devices that will weaponize an agent prior to or during the activity. Simple, timed aerosol devices can release agent at a specific moment, and serve as low-technology weapons that result in large numbers of initial victims. Over subsequent hours, the devices could fell more and more victims, creating a large-scale MCI.

(For further information see **Chapter VIII**, Entertainment Facilities)

Special Events

Emergency response officials have also closely studied the potential for terrorist attacks at extraordinary and/or rare celebrations, such as the Olympic Games, New Year's Eve celebrations and independence day spectaculars.

Security planning surrounding the 1996 Atlanta and the 2000 Sydney Olympic Games focused on possible terrorist activity at high-publicity special events.

■ Despite limited casualties, the Olympic Centennial Park bombing in Atlanta in 1996 illustrates the threat. The bomber placed a device in an area frequented by spectators, but which also had a limited security presence. Subsequent media attention devoted to the Olympic bombing helped the terrorists achieve their goals, by demonstrating the vulnerability of the US to terrorism.

(For further information see **Chapter IX,** Special Events)

Chapter II: Security Planning

Facility Security Planning

Introduction

Performing a Security Survey

Planning for Bomb Threats and Incidents

Search Teams

Anti-Terrorism Survey

Workplace Security Survey

Facility Security Planning

INTRODUCTION

It is difficult to overstate the importance of proactive security measures to combat terrorism. Terrorists frequently target sites that offer ease of access.

Businesses need to be open to the public to maintain favorable relations with consumers. Any increase in security measures that restricts freedom of access can unavoidably damage the **corporate image** to some extent. Encumbering security procedures have the potential to put off current or prospective clients. **Governments** also must weigh the consequences of appearing to be frightened.

Yet many industries have already suffered some sort of violation — whether of a **criminal** or **terrorist** nature. In hindsight, many international air carriers, banks, energy companies, insurance agencies and retail companies have wondered why they were not prepared for a terrorist situation. The reason is simply that security is a low priority in most corporations and agencies. Most believe that terrorism happens elsewhere, or that costs associated with preventive measures are too great.

The key to a successful security program is a well-planned and coordinated team effort, that uses timely, accurate information to respond to terrorists.

Performing A Security Survey

A security survey is a complete inspection and analysis of a facility, intended to uncover security flaws and assess facility security procedures and equipment.

No one standard checklist can measure the security needs of every company, institution or facility type. Each organization needs to conduct a survey **specific to its daily operations**.

Security surveys should include on-site evaluation. In order to be truly effective, a trained crime prevention officer, security manager or private security consultant should conduct a physical security survey of the facility.

Security Survey Objectives
A security survey should have three principal objectives. If met, the survey should help to prevent a terrorist attack from occurring. Planners should:

1. Assess a facility's present security status.
2. Identify existing security deficiencies.
3. Suggest improvements to make it more difficult for terrorists to commit an act against a facility.

Security personnel can accomplish the first two objectives by compiling a detailed analysis of current external and internal security activities.

■ *Typical inspections cover the surrounding area, security lighting, doors, locks, windows, alarm systems as well as structural and environmental designs.*

Security surveys should also include an evaluation of current operating policies and procedures at a facility. For example, in a large office complex, a security review should include an analysis of shipping, receiving, contractor permits, parking permits and key access to ensure proper controls.

Planners should produce specific recommendations based on the survey results, regarding increased security measures and the costs associated with each improvement.

Planning a Survey

A survey should be a well-thought out and planned process, involving an evaluation of the security measures required to protect individuals, families, fellow workers and assets. Planners should compile and file detailed lists of individuals who have responsibility for specific security areas, along with contact information. These people will gather and evaluate specific information concerning an area.

Elements of the Survey

The three major elements of a security survey include (1) **physical security**, (2) **personnel security** and (3) **information security**. Assessing physical security involves the inspection of all structural, physical, environmental and architectural aspects of the facility. This assessment includes doors, windows, locks, lighting, fencing, alarms and facility location.

Personnel security assessments treat threats an individual might expect to encounter from an evaluation of the security needs of a family with children, to a staff of employers and executives.

Information security involves an assessment of the level of control of all forms of printed material and communications (verbal, fax and data), as well as computer system/network security. Information security also involves all the records, documents, correspondence and/or vital plans of a corporation.

Survey Instruments
Due to vast differences in companies, buildings and individuals, many security survey instruments and methods can be valid. Surveys range from a standard checklist to a 100-page report, but no matter the method, the report should remain confidential and planners must destroy old reports when outdated.

■ *In the wrong hands a survey report can illustrate in detail facility vulnerabilities and serve as an invitation for a terrorist to attack.*

Profile of a Security Surveyor

The individual who conducts a survey should be reasonably suspicious, alert and able to make deductive leaps. Local and federal law enforcement agencies frequently conduct surveys as a service to the community. Federal and state agencies can and often do assist in comprehensive surveys, but budgetary constraints and/or a lack of manpower have tended to reduce the extent of this assistance.

Planning for Bomb Threats and Incidents

The threat of a bomb attack is a reality in today's world. The best defense against bombers is simply adequate preparation — a bombing should never catch businesses or institutions off-guard. A **bomb incident plan** within a physical security plan can greatly reduce the risk of casualties and damage.

Bomb Threats and Bombs

Terrorists can deliver bomb threats in a variety of ways, though the majority of such threats are today received by telephone. Bomb threats originate either from the individual who placed a device, or occasionally by someone with knowledge of a device who may wish to minimize personal injury or property damage.

Terrorists may place these calls to a third party, such as a news agency or radio station, else they may communicate in writing or by recording. The threat may be false, for the caller may want to create an atmosphere of panic to disrupt normal activities at a facility. Whatever the reason behind a threat, security personnel *must* react.

Terrorists can construct bombs that look like almost anything, and deliver devices in a number of ways. Few devices actually resemble stereotypical bombs. Most bombs are home-made, and are limited in design only by the imagination and resources of the bomber.

■ *Suspect anything that looks unusual when searching for bombs.*

Preparing and Planning for a Bomb Incident

The creation and employment of security measures can **harden** a target to terrorist attack. In a bomb incident, proper planning will reduce the likelihood of panic and instill public confidence in authorities.

To prepare for a bomb incident, develop separate but interdependent plans: a **bomb incident plan** and a **physical security plan**.

Bomb incident plans detail procedures for implementation when terrorists execute or threaten to execute an attack. Planners should establish a clear **chain of command** or **line of authority** for bomb events. A chain of command is simple if there is already an office structure; e.g., one business occupies one building. In complex situations (such as multi-occupant buildings), send representatives from each business/institution to attend planning seminars. Appoint a seminar leader and delineate a clear line of succession.

Designate a command center or other focal point for telephone and radio communications in the planning phase; permit only individuals with assigned duties in the center. Plan for alternates in case someone is absent in the event a threat arrives.

■ **Establish lines of communication, between the command center and search and evacuation teams. Maintain an updated blueprint or floor plan of the building in the command center.**

Responding to Bomb Threats

All personnel require training to respond effectively to a bomb threat, but telephone personnel require particular emphasis. More than one person should listen in on a call.

■ **Develop a covert signaling system to notify others of a called bomb threat. Calm responses may elicit additional information, especially if the caller wishes to avoid injuries or deaths.**

Keep in mind that the bomb threat caller is almost always the best source of information about the bomb. Use a standard bomb threat checklist during the bomb threat call to gather important information regarding the bomb.

The checklist will later help law enforcement officials investigate the incident and assist in the search for the device. The receptionists' responses should aim at keeping the caller on the line as long as possible. The operator should ask him/her to repeat the message and should write down every word spoken. If the caller does not indicate the location of the bomb or the time of detonation, then the operator should request this information. The person receiving the threat should inform the caller that the building is occupied and that death or serious injury to innocents could result.

■ *Pay close attention to background noises, such as running motors and music, which can provide clues about the caller's location. The operator should listen closely to the voice, its tone, accent and/or any speech impediments.*

Immediately after the caller hangs up, the operator should report the threat to the person designated to handle facility security, and immediately contact the police department, fire department and other agencies. The **bomb incident plan** should have established the sequence of notification.

Finally, the operator should remain available for interviews with law enforcement personnel. If the bomb threat is in writing, save materials such as envelopes or containers. Avoid further handling of materials after determining that the message is a bomb threat. Preserve all evidence, including fingerprints, paper, postal marks, handwriting or typewriting.

Evacuation Decision

The most weighty management decision in the event of a bomb threat is whether or not to evacuate the building. The bomb incident plan may already specify when to make that decision. An evacuation decision demonstrates a deep concern for the safety of personnel in the building, but it also results in a costly loss of productive time.

There are three alternatives when faced with a bomb threat. The first is to **ignore the threat**, the second is to **evacuate immediately** and the third is to **conduct a search** and then to **evacuate** if the situation so warrants.

Ignoring the threat completely can cause a number of problems. While few bomb threats are real, some do lead to real devices. If employees learn that the institution/business has received and ignored bomb threats, there could be a deleterious effect on morale over the long-term.

■ *If callers feel security planners are ignoring their threats, they may launch real attacks.*

Evacuating immediately on every bomb threat seems a wise approach at first glance. Immediate evacuation causes a disruption of work. Should terrorists learn that building policy dictates evacuation with each threat, continuous calls could force business to a standstill. Moreover, employees can call in a threat to avoid work, or students to miss class or exams.

Organize and train an evacuation team in coordination with bomb incident plan development. Police and fire departments within the community can usually provide this type of evacuation training.

■ *Search personnel must be intimately familiar with every location in the building where a device might be concealed. Police officers or fire-fighters will be unfamiliar with the location.*

The execution of a search after the receipt of a threat is the third, and perhaps the best approach. This theory proposes that personnel only evacuate a facility following the actual discovery of a suspicious package or device. This may be less disruptive than immediate evacuations.

SEARCH TEAMS

More than one individual should search each area. Supervisory personnel can conduct rapid searches without causing disturbances, but supervisory searches may be inadequate, due to unfamiliarity with areas. Area occupants can search their own areas, as they are most concerned for their own safety. They are also familiar with what does or does not belong in their areas.

■ *Searches by well-trained teams are best for safety, morale and thoroughness, although this does take the most time.*

The use of trained search teams will significantly impact on productivity. Bomb searches are slow operations that require comprehensive training and practice. Management must determine who should conduct searches, and should be heavily involved in the creation of bomb incident plans.

Search Technique
The composition of search teams will vary with each facility and its size, but for the purpose of this example a two-person search team will suffice.

When a two-person search team enters a room, they should first move to the central part of the room, stand quietly with their eyes closed, and **listen for a clockwork (ticking) device**. Often, no special equipment is needed to detect a clockwork mechanism. Even if there is no ticking sound, the team members will make themselves aware of the level of background noise within the room.

Background noise or transferred sound is always present in building searches. Unidentified ticking sounds may be clearly audible but may not be found, and can become a source of distraction and unease. Ticking sounds can emanate from unbalanced air-conditioner fans floors away, or alternatively from a dripping sink down a hall. Sound can travel through air-conditioning ducts, through water pipes and walls. Background noise can include outside traffic sounds, rain and wind.

The leader of a two-person search team should **divide the room** into two parts. This division is based on the number and type of objects in the room, not its dimensions. Draw an imaginary line between two objects (e.g. the edge of a window on the north wall to the picture on a south wall).

First Room Sweep

■ *The first sweep covers items that rest on the floor up to a certain height. Furniture or other objects in the room can help to determine this height, although it is usually hip level.*

Both individuals start at each end of the room division line, from a back-to-back position. This is the starting point for each successive search sweep. The search team works toward each other, inspecting the wall area and all items on the floor. When the team members meet, they will have completed a **wall sweep**.

The team should check all items in the middle of the room up to the selected level, **including the floor under the carpeting**. This first search sweep should also cover items mounted on or in the walls, such as air-conditioning ducts, baseboard heaters and wall units, if below hip height.

Second Room Sweep

■ *The second sweep height usually extends from the hip to the chin or top of the head.*

This sweep usually covers pictures hanging on the walls, built-in bookcases and tall light fixtures.

Third Room Sweep

■ *The third sweep covers the area from the chin or top of the head to the ceiling.*

The sweep covers high mounted air-conditioning ducts and hanging light fixtures.

Fourth Room Sweep

■ *If the room has a false or suspended ceiling, the fourth sweep investigates this level.*

Check flush or ceiling-mounted light fixtures, air-conditioning or ventilation ducts, structural frame members, electrical wiring, and sound or speaker systems.

Upon completion of the search, post a sign or marker that says **'Search Completed'** on the entrance door. If the use of signs is not practical, colored tape across the door will suffice.

Similar search techniques can serve for any enclosed area or the exterior of a building, with modifications where appropriate. For example, if there is a threat on the life of a guest speaker at a convention, thoroughly search the platform and microphones first, but always return to the searching technique. **Random or spot-checking of only apparently logical areas is insufficient**.

Locating Suspicious Objects

It is imperative that search teams know that their *only* mission is to search out and report discoveries of any suspicious objects.

■ *Under no circumstances should search team members move or touch suspicious objects. Leave the removal or disarmament of bombs to professionals with the bomb disposal unit.*

If searches yield a suspicious object:

1. Report the location and an accurate description of the object. Relay this information immediately to the command center, which should then notify the police and fire departments. Meet and escort these officers to the scene.

2. If it is absolutely necessary, place sandbags or mattresses — never metal shields — around any suspicious objects. Do not attempt to cover objects.

3. Identify the danger area and block off a clear zone of at least 300 feet, including floors directly below and above the object.

4. Check to see that all doors and windows are open. This will minimize primary and/or secondary blast damage from fragmentation.

5. Evacuate the building.

6. Do not permit re-entry until after professional removal/disarmament of the device, and a declaration that the building is safe for re-entry.

ANTI-TERRORISM SURVEY

The following survey can help emergency response personnel perform a self-assessment of their daily activities, and is derived from the 1996 Anti-terrorism Training Program of the Commonwealth of Massachusetts, and from Arthur A. Kingsbury, *Introduction to Security and Crime Prevention Surveys*. This survey can help security planners identify security weaknesses in the workplace and at home. It can also comprise part of a crime prevention awareness program with the public.

Personal Security Survey

Yes/No **Do you. . .**

☐ ☐ 1. Have the names and identification numbers of all your credit cards written down and kept in a safe place?

☐ ☐ 2. Maintain a 'no-fight' policy if robbed at gun point or knife point?

☐ ☐ 3. Have a cellular or portable telephone in a vehicle?

☐ ☐ 4. Always lock the car when exiting?

☐ ☐ 5. Park in well-lit areas?

☐ ☐ 6. Check the car's safety equipment frequently and always keep the gas tank one fourth to one half full?

☐ ☐ 7. Always drive with the windows up and the doors locked?

Yes/No **Do you. . .**

☐ ☐ 8. Look inside your vehicle upon approaching and entering it?

☐ ☐ 9. Conduct a check around the vehicle before approaching it?

☐ ☐ 10. Feel you should quicken your pace or run if you suspect you are being followed?

☐ ☐ 11. Avoid carrying keys with attached identification and markings?

☐ ☐ 12. Carry only the minimum amount of cash needed?

☐ ☐ 13. Avoid appearing flashy and flamboyant?

☐ ☐ 14. Usually go out shopping with someone else?

☐ ☐ 15. Have keys ready when approaching the car or home door?

☐ ☐ 16. Avoid discussing income and personal business with anyone?

☐ ☐ 17. Avoid unnecessary night trips?

☐ ☐ 18. Refrain from taking short cuts through vacant lots, alleys, etc.?

☐ ☐ 19. Refrain from picking up hitchhikers or stopping for stalled cars, no matter what the circumstances?

☐ ☐ 20. Refrain from hitchhiking?

☐ ☐ 21. Know your associates and inform others when going out?

☐ ☐ 22. Screen and check domestic employees' references?

Yes/No **Do you. . .**

❏ ❏ 23. Inform babysitters of your location and how to contact you in case of an emergency?

❏ ❏ 24. Supervise workmen or servicemen while they are working on the home or office?

❏ ❏ 25. Instruct children in personnel safety matters, particularly if they walk to and from school?

❏ ❏ 26. Know how to respond to obscene or harassing telephone calls?

❏ ❏ 27. Have an unlisted telephone number?

Comments:

WORKPLACE SECURITY SURVEY

Access Control

Yes/No

☐ ☐ 1. Do visitors need to secure passes before they enter?

☐ ☐ 2. Are visitor passes distinctive from employee passes?

☐ ☐ 3. Is there a record of when and to whom the organization issues passes?

☐ ☐ 4. Does the organization collect passes when visitors depart?

☐ ☐ 5. Are passes or badges difficult to forge?

☐ ☐ 6. Is the perimeter of the office or building adequately illuminated?

☐ ☐ 7. Is the roof illuminated?

☐ ☐ 8. Are the parking lots adequately illuminated?

☐ ☐ 9. Do time-sensitive or motion sensor devices control the lights?

☐ ☐ 10. Does the organization replace burnt-out light bulbs immediately?

☐ ☐ 11. Are light fixtures protected against breakage?

☐ ☐ 12. Are passageways and storage areas illuminated?

☐ ☐ 13. Is lighting at night sufficient for police surveillance?

Yes/No

❏ ❏ 14. Does a fence or wall protect the place of business on all sides?

❏ ❏ 15. Are fences or walls in good repair?

❏ ❏ 16. Do groundskeepers keep the fence or wall clear of nearby trees, bushes and tall grass?

❏ ❏ 17. Does Security check locks regularly?

❏ ❏ 18. Do gates remain locked when not in use?

❏ ❏ 19. Is there an alarm system?

❏ ❏ 20. Are there security locking devices on each door and window?

❏ ❏ 21. Are doors constructed of sturdy materials?

❏ ❏ 22. Are there only the barest minimum of access doors to the facility?

❏ ❏ 23. Are door hinges spot-welded or secured, in order to prevent removal?

❏ ❏ 24. Are the hinges facing the inward side of the doors?

❏ ❏ 25. Are there time locks to detect unauthorized entrance?

❏ ❏ 26. If there are padlocks, do they comprise high-quality materials?

❏ ❏ 27. Are padlock hasps made of heavy duty materials?

❏ ❏ 28. Do opening alarms protect all fire doors?

❏ ❏ 29. Is the alarm system connected to all doors and windows?

Yes/No
- ❏ ❏ 30. Does the organization follow a specific lock-up procedure?
- ❏ ❏ 31. Is someone responsible for checking all doors and windows to make sure they are closed and locked every night?
- ❏ ❏ 32. Are all alarms connected to a central control center?
- ❏ ❏ 33. Do personnel man the station at all times?
- ❏ ❏ 34. Are there periodic checks on response times to alarms?
- ❏ ❏ 35. Does the organization test alarms on a regular basis?
- ❏ ❏ 36. Is there a backup emergency power source for the alarm system?
- ❏ ❏ 37. Are surveillance cameras in place for all exits and entrances?
- ❏ ❏ 38. Are surveillance cameras in place for all parking lots and alleys?

Vehicle Control

Yes/No
- ❏ ❏ 39. Is there a separate area for employee parking?
- ❏ ❏ 40. Is there a separate area for visitor parking?
- ❏ ❏ 41. Do personnel verify all service vehicles?
- ❏ ❏ 42. Is there a log of service vehicles?
- ❏ ❏ 43. Does the organization fence in or secure parking areas?

Yes/No
☐ ☐ 44. Does the organization illuminate parking areas?
☐ ☐ 45. Do guards patrol parking areas?

Office Security
Yes/No
☐ ☐ 1. Do personnel properly greet and/or challenge strangers?
☐ ☐ 2. Do personnel protect billfolds, purses and other personal belongings while on the job?
☐ ☐ 3. Does only one person issue all keys?
☐ ☐ 4. Does the organization keep a record of who has received what keys, and if the individual(s) return them?
☐ ☐ 5. Do all keys clearly state "Do Not Duplicate?"
☐ ☐ 6. Does the organization have a lost key policy?
☐ ☐ 7. Are maintenance personnel, visitors, etc. required to show ID to a receptionist?
☐ ☐ 8. Is there a clear line of sight from the reception area to the entrance, stairs and elevators?
☐ ☐ 9. Is it possible to reduce the number of entrances without a loss of efficiency or safety?
☐ ☐ 10. Do personnel keep office doors locked when unattended for a long period of time?

Yes/No

☐ ☐ 11. Do personnel keep items of value secure in a locked file or desk drawer?

☐ ☐ 12. Has Security briefed the supervisor of each office on security problems and procedures?

☐ ☐ 13. Do all office employees receive some security education?

☐ ☐ 14. Do office-closing procedures require that important information be secure at night?

☐ ☐ 15. Does the organization keep office entrance doors locked except during business hours?

☐ ☐ 16. Do personnel shred confidential material before placing it in the trash?

☐ ☐ 17. Does the organization log in and out all janitorial and cleaning services personnel?

☐ ☐ 18. Does a security force protect the facility or building?

☐ ☐ 19. Do guards understand their role?

☐ ☐ 20. Are guards prepared to act in case of an emergency?

☐ ☐ 21. Do guards carry arms legally?

☐ ☐ 22. Are guards alert?

☐ ☐ 23. Is there an effective system of communication for emergency situations?

High Security Areas
Yes/No
- ☐ ☐ 24. Do personnel keep high security areas locked at all times?
- ☐ ☐ 25. Do managers and/or security personnel keep high security areas under supervision?
- ☐ ☐ 26. Do badges bear clear markings to designate those who may enter security areas?
- ☐ ☐ 27. Do procedures require employees to verify their identity when entering security areas?
- ☐ ☐ 28. Is access to high security areas controlled?

Personnel
Yes/No **Does Security...**
- ☐ ☐ 29. Require personnel to wear badges or identification cards?
- ☐ ☐ 30. Require employees to display ID badges at entrances?
- ☐ ☐ 31. Include numbers on all identification cards?
- ☐ ☐ 32. Include employee photographs on all ID cards?
- ☐ ☐ 33. Keep a record of all lost or stolen badges?
- ☐ ☐ 34. Keep a record of all badges issued?
- ☐ ☐ 35. Institute standard screening procedures for all employees before hiring?

Yes/No

☐ ☐ 36. Fingerprint all employees?

☐ ☐ 37. Photograph all applicants?

☐ ☐ 38. Keep personnel files of all employees?

☐ ☐ 39. Require employees to produce official identification at the time of hiring?

☐ ☐ 40. Check references?

☐ ☐ 41. Require employees to present a list of past employees?

☐ ☐ 42. Check employees' past employers?

☐ ☐ 43. Require employees to provide any pseudonyms?

☐ ☐ 44. Instruct employees on all security and emergency operating procedures in place?

Comments:

Apartment Security Survey

Yes/No

☐ ☐ 1. Is there adequate interior and exterior lighting?

☐ ☐ 2. Can residents eliminate blind spots or hiding places around the apartment?

☐ ☐ 3. Does the apartment have a doorman and/or security guard?

☐ ☐ 4. Does the doorman and/or security guard screen visitors?

☐ ☐ 5. Do cameras and personnel monitor elevators?

☐ ☐ 6. Is there adequate protection for apartment windows?

☐ ☐ 7. Do neighbors know each others' names and phone numbers?

☐ ☐ 8. Did residents have the locks changed when they moved into the apartment?

☐ ☐ 9. Do residents report anything out of order and institute repairs immediately?

☐ ☐ 10. Do residents avoid trips to the laundry room or mail box late at night?

☐ ☐ 11. Do residents avoid admitting persons into the building unless they know the visitors' identity and purpose?

☐ ☐ 12. Do residents use only their first initials and last names on mailboxes and telephone listings — especially if female and living alone?

☐ ☐ 13. Does the apartment complex have a supervised playground area for children?

☐ ☐ 14. Is the parking area secure?

Comments:

Home Security Survey

Yes/No

☐ ☐ 1. Are entrance doors of a solid core type?
☐ ☐ 2. Do they have deadbolt locks?
☐ ☐ 3. Do bolts extend at least three-fourths of an inch into the strike?
☐ ☐ 4. Is there little or no 'play' when you try to force the door bolt out of the strike by prying the door away from the frame?
☐ ☐ 5. Are doors in good repair?
☐ ☐ 6. Are locks firmly mounted?
☐ ☐ 7. Are all doors securely mounted?
☐ ☐ 8. Can an assailant open any door by breaking a window or a panel of wood?
☐ ☐ 9. Has the management permanently secured any unused doors?
☐ ☐ 10. Are roof hatches, trap doors or roof doors properly secure?
☐ ☐ 11. Are there adequate locks on bedroom doors?
☐ ☐ 12. Are exterior doors generally locked?
☐ ☐ 13. Does the front door have a 180° view peep hole?
☐ ☐ 14. Can visitors enter before confirmation of their identity and the purpose of their visit?
☐ ☐ 15. Do residents check and verify unsolicited callers before they allow callers to enter?
☐ ☐ 16. Do patio doors have adequate locks?

❏ ❏ 17. Are garage doors locked at all times, particularly at night and when residents are away?

❏ ❏ 18. Do residents use automatic door openers?

❏ ❏ 19. Do residents change electronic codes frequently?

❏ ❏ 20. Are there adequate locks on tool sheds, greenhouses and similar structures?

❏ ❏ 21. Do residents avoid keeping a key 'hidden' outside the home?

❏ ❏ 22. Do residents keep windows locked at all times?

❏ ❏ 23. Are window and wall air conditioners secured against removal?

❏ ❏ 24. Have residents removed ladders, trellises or similar climbing aids to prevent entry into the second story window?

❏ ❏ 25. Is indoor lighting functional?

❏ ❏ 26. Is outdoor lighting adequate?

❏ ❏ 27. Do lights illuminate the sides of the residence and garage area?

❏ ❏ 28. Do residents leave lights on during all hours of darkness?

❏ ❏ 29. Do residents turn on outside lights on before they leave the house at night?

❏ ❏ 30. Do residents report broken streetlights immediately?

❏ ❏ 31. Do fences protect the property or do they provide a hiding place for an intruder?

❐ ❐ 32. Are gates in good repair and have they locks?

❐ ❐ 33. Is there a watchdog or other family pet?

❐ ❐ 34. Do residents belong to a neighborhood watch program?

❐ ❐ 35. Do residents draw drapes or shades at night?

❐ ❐ 36. Do family members keep an alert watch on persons who may be surveilling or casing the home?

❐ ❐ 37. Can residents lock the mailbox?

❐ ❐ 38. Do residents turn on lights and make noise if awakened during the night?

❐ ❐ 39. Do residents have an alarm system with a panic button?

❐ ❐ 40. Do residents keep a flashlight by the bed?

❐ ❐ 41. Is there a telephone in the bedroom?

❐ ❐ 42. Do residents have a list of the neighbors' telephone numbers?

❐ ❐ 43. Do residents have the numbers to the police, fire and EMS by the telephone?

❐ ❐ 44. Do neighbors have residents' telephone number?

Comments:

Chapter III: Buildings Security

Threats to Buildings from Terrorist Attack

Introduction

Likely Target Structures

Terrorist Threats to Building Security

Why Buildings are Vulnerable to Terrorist Attack

Vulnerability Assessment: Building Profile

Vulnerability Assessment Tools and Techniques

Hardening a Target

Securing Access and Entry Points

Threats to Buildings from Terrorist Attack

INTRODUCTION

An overwhelming number of institutions/businesses concerned with terrorism reside in single buildings, and not expansive complexes or campuses. Protecting one building can be an easier undertaking than the protection of a complex or campus, but this does not diminish the risk from attack.

One pressing counter-terrorism concern is the prospect of **chemical-biological (CB) agent delivery** as an aerosol into a closed system. CB agents are concealable, travel quickly through ventilation systems, and can disseminate rapidly throughout a building, with the potential to produce high mortality rates.

■ *Knowledge of building systems and dynamics is critical in order to evaluate risk areas and prevent entry and transport of CB agents into a facility.*

Virtually any building is a potential target. However, specific profiles can identify buildings that are most vulnerable. Management and occupants should treat the threat of terrorist incidents as they would address many other emergency concerns.

Using a checklist format, an environmental security plan can build upon procedures already in place to deal with bomb threats, fire evacuations, and the handling of biomedical and chemical waste.

◆ The first step to protect against a terrorist threat is the formulation of a thorough vulnerability analysis.
◆ Then take steps to harden the facility.

LIKELY TARGET STRUCTURES

Every building and facility is a potential target, but three basic building types are most vulnerable to attack:

◆ Buildings that house companies or organizations.
◆ Buildings that, if damaged, would disturb the public, the economy, or have the potential to provoke mass hysteria.
◆ Buildings that possess symbolic value.

Buildings that house companies or organizations

These may be at risk from disgruntled employees, customers or other aggrieved parties. This definition includes both irate employees and external populations fighting for a social or political cause. Likely targets might manufacture controversial products, support controversial company/government policies or actions, or render financial assistance to certain organizations or countries.

Examples: banks, chemical or drug companies, government offices or hospitals and medical clinics.

Buildings that, if damaged, would disturb the public or the economy, or provoke mass hysteria

The goal in attacks on such structures is to generate confusion and chaos within a geographical area, and thereby capture national attention and headlines.

Examples: airports, train stations, sports stadiums, skyscrapers and financial exchanges.

Buildings that possess symbolic value

Attackers of such buildings seek to exact revenge or incite hatred. Nationally and internationally recognized buildings are always vulnerable to terrorist attack.

Examples: domestic or multinational monuments and institutions, government buildings, corporate facilities, airports and facilities that house large events.

TERRORIST THREATS TO BUILDING SECURITY

The Exterior Threat Environment

The first step is determining the threat level in the area. A comprehensive threat assessment will determine the degree of caution and amount of protection planners should employ to prevent a terrorist confrontation.

Determining the threat level is not an especially difficult process. **Open source information** exists and can aid in the assessment process. Another source of information can be the development of an **interagency working group**, including public and private agencies concerned with terrorist threats in the given area. The guidelines in Table I are intended to assist in threat assessment.

Physical security is designed to protect property, personnel and material against attack. Proactive security measures can address security needs to make an individual or facility harder for a terrorist to attack.

■ *Fences and heavy barriers at entrances and around facilities can be effective target-hardening measures.*

Table I: Threat Assessment Guidelines	
Characteristic	**Consequence**
Existence	A terrorist group is present in the area or could gain access into a country or group.
Credibility	The ability to carry out an attack has been assessed and demonstrated.
Intent	Evidence of demonstrated terrorist activity, threat or action by a group
History	Demonstrated terrorist activity over time.
Targeting	Current credible information exists on activities indicative of preparations for specific terrorist operations.
Security Information	The internal politics and security considerations that impact on the capabilities of the terrorist element to carry out its mission.

■ *Related security measures include security patrols or other hi-tech security enhancements.*

Soft targets — those targets not adequately concerned with security — are easily accessible, betray predictable patterns in their daily routine and are generally not conscious of security.

Security Measures
Typical security measures include detection systems, barriers, panic alarms, uniformed guards and

high visibility security patrols. The desire to commit a terrorist act can be significantly decreased through simple increases in security measures.

■ *Outside lights and locks on windows and doors are designed and installed to contribute to facility security and the protection of its occupants.*

The exterior configuration of a building or facility is crucial. In most instances architects give little or no consideration to security concerns, but security personnel can take steps to offset design oversight.

◆ Fencing, lighting and controlled access can reduce facility vulnerability to terrorist attack considerably.
◆ Parking should be restricted if possible to 300 feet from the building or any building in a complex. If restricted parking is not feasible, only properly identified employee vehicles should be immediately next to the facility; visitor vehicles should park at a distance.
◆ Trim heavy shrubs and vines close to the ground to reduce their ability to conceal criminals or bombs.
◆ Unless there is an absolute requirement for such ornamentation, remove window boxes and planters, as they are perfect receptacles for bombs. If they must remain, ensure that security patrols check such receptacles regularly.

Install entrance/exit doors with the hinges and hinge pins located on the interior, to prevent their removal. **Solid wood** or **sheet metal-faced doors**, preferably in a steel door frame, provide additional integrity than that afforded by a **hollow-core wooden door**.

The ideal security situation is a building without **windows**. However, bars, grates, heavy mesh screens or steel shutters over windows can offer significant protection from unwanted entry.

Cover **floor vents** and ensure that openings in the protective coverings are not too large. Establish **access controls**, to identify personnel authorized to enter critical areas, and to deny access to unauthorized personnel. Apply these controls to the inspection of all packages and materials that enter critical areas.

Special Glass
The development of special glass in recent years has enhanced security effectiveness in a number of ways. Mylar film and polymer glass can decrease the danger from flying glass in case of exterior impact, or in the event of an explosion from within the facility.

■ *Ballistic treatments render some glass bulletproof, while wire reinforcements can prevent the penetration of large objects such as rocks, pipe bombs or grenades through broken glass. This type of glass should be used on the bottom floors of buildings — where they are most vulnerable.*

Barriers

Fences and barriers against intruders are common at many installations and facilities, although most such barriers are intended to stop petty thieves, not well-armed and determined terrorists.

Vision Barriers

The use of vision barriers can prevent potential terrorist attacks, since they hinder invaluable surveillance activity. This benefit is nullified, however, if the facility routinely permits unscreened random visitors to tour the facility. This enables terrorists to perform unrestricted surveillance of an otherwise hidden area.

Fences

Chain-link steel fences, topped with razor wire, are common perimeter barriers. When backed up with a conscientious guard force, fencing can be effective against casual or determined intruders.

■ *Holes or unlocked gates render access to a building easy. Determined intruders can use wire cutters, ladders or a vehicle to circumvent fencing.*

Still, a well-maintained, lighted and patrolled perimeter fence can be the first line in a solid defense. Well-patrolled and monitored with **CCTV**, fencing can be very effective against attempts to breach and/or reconnoiter a facility.

Heavy Barriers

A truck carrying explosives can damage or destroy important support structures if permitted to gain access to a building or structure. **Heavy jersey barriers** or **staggered cement flower pots** can delay or even defeat such attacks.

The Indoor Environment — A Major Terrorist Target

Different definitions of the term "environment" can assist in the identification of targets for domestic counter-terrorism assessments.

In general terms, the environment is all that surrounds us. There is the **natural environment**, which includes the outdoors or ambient environment. Then there is the **indoor** or **microenvironment** created by humans, the "built environment."

The built environment includes the outside envelope (e.g., exterior walls, loading dock, ramps, etc.), as well as the interior of a structure.

■ *These structures include not only office buildings, but also train stations, airports, ship terminals, sports stadiums, and public buildings such as monuments or memorials.*

WHY BUILDINGS ARE VULNERABLE TO TERRORIST ATTACK

Building Structures and Systems

Buildings are particularly vulnerable to Chemical-Biological (CB) attack due to ease of access to building systems. All buildings have **entry and access points** that allow for occupant entry and transport. They also contain **utility and environmental systems** to support life and its activities.

Buildings are made up of several basic structural components and systems. **Structural components** include the foundation, walls, domes and vaults, arches, columns and beams of the building. **Building systems** comprise standards and codes, heating systems, ventilation systems, air conditioning systems, plumbing systems, electrical systems, fire protection systems and communication systems.

Structural Components

◆ The *foundation* distributes weight load by increasing the load bearing area. Using a broad foundation can reduce the stress and strain on the load bearing soil(s) beneath the building.

◆ *Walls* can bear weight and also add stability if placed as right-angled connections.

◆ *Columns* allow a structure to be opened up while still safely distributing the weight load down to lower levels and the foundation.

◆ *Arches* distribute loads sideways to the base, due to the shape of stones. However, the upward thrust at the base must be contained (built up) or the arch may fail.

◆ *Domes* and *Vaults* can be viewed as rotated arches.

◆ *Beams* span the load bearing structures (walls and columns) and support the load of building contents.

Timber joists are simple wood beams that span short lengths of residential (low-rise) construction.

I-Beams (universal beams) are long steel beams whose design provides great resistance to stress. This design permits a beam to support equal or greater loads than a solid joist of much greater weight.

Concrete reinforced beams take advantage of the strengths of two materials. Concrete withstands compression well, but requires the addition of steel rods (rebar) to withstand tension.

Building Shells

There are two basic structure types, or shells, that fit the profile of a potential terrorist target.

◆ *High-rise cellular* construction is a box type construction, with solid walls in a fixed, repeated pattern, supporting the load to the foundation.

It is used in residential, commercial and industrial buildings, and consists of a concrete foundation, a frame of brick with wood joists, walls (comprising solid concrete, insulation and interior masonry), as well as a roof of flat steel and concrete with a built-up composite surface. Cellular construction is limited to a height of about 15 stories.

◆ *High-rise skeleton* construction is a skyscraper construction of steel and concrete, 20 or more stories high.

The combined strength of steel and concrete allows for large, open floor areas without the need for cellular box type construction. As the building height increases, trussing may be required.

This type of construction appears in residential, commercial and industrial buildings. It consists of a foundation made of concrete floor slab, a frame of steel beams, concrete walls with insulation and interior masonry, a roof of flat steel and concrete with a built-up composite surface.

Building Systems
Complex, interrelated systems work together to ensure safety, comfort and convenience. These systems include: standards and codes, heating, ventilation, air conditioning, plumbing, electrical, fire protection, and communication.

◆ *Heating systems* include steam, hot water and forced air. They consist of a fuel source, heat generating equipment and heat distribution systems. **Fuel sources** are wood, fossil fuels (coal, oil, gas) and electricity. **Heat generating equipment** includes boilers (hot water and steam), furnaces (forced air), heat exchangers (hot water), utility-generated (steam) and resistance heaters (electricity). **Heat distribution systems** are comprised of hot water piping, steam piping, ductwork, pumps, fans and radiant systems. Examples of heating controls include thermostats and safety controls.

◆ *Ventilation systems* remove stale air, odors and particles from building spaces and introduce fresh, clean air. Most ventilation systems consist of ductwork and ventilating equipment. The ductwork has air supply ducts (in) and air exhaust ducts (out). The ventilating equipment consists of fans, registers, dampers and controls.

◆ *Air conditioning systems* cool inside air, regulate humidity and clean, change and circulate air. There are a variety of cooling methods, which may use all air; air and water; all water; or a direct refrigerant.

◆ *Plumbing systems* provide potable water for human consumption, water for heating and cooling systems and also water for the sanitary removal of human waste. The major components of plumbing systems include pipes, fixtures and fittings. Piping is used for potable water, hot water, steam delivery, and also sanitary wastewater and storm water transferal. Other equipment used in plumbing systems are boilers, hot water heaters, heating coils and roof storage tanks.

◆ *Electrical systems* bring in high voltage power from utility companies and then condition and distribute usable power throughout the building. Systems include the utility feed, service switch, panel boards, cables, conduits, raceways, circuit breakers, switches and outlets.

◆ *Fire protection systems* provide dedicated fire protection and fire suppression equipment. Fire protection systems include fire-resistant building materials and furnishings, fire doors and other airflow barriers, well-marked exits, electrical system safeguards, smoke, heat and gas detection equipment, standpipe and hose systems, sprinkler systems, and extinguishers.

◆ *Communication systems* manage the movement of information to and from the building, and within the building. Information can travel in the form of voice communication (telephone) or data communication (computer, fax, modem) and may travel over conventional copper wire (twisted pair, coaxial) or fiberoptic cables (glass).

Entry Points for Terrorist CB Attack

◆ Intentional Air Intakes — Entry and Access Points.

◆ Unintentional Air Intakes — Entry and Access Points.

◆ Heating, Ventilation and Air-conditioning Systems (HVAC).

While building systems shield occupants and their activities from outside elements, these same systems also provide opportunities for terrorists to release CB agents.

There are numerous access and entry points for CB agent dispersion throughout building pathways — intakes labeled either **intentional** or **unintentional** in industry jargon.

◆ *Intentional Air Intakes — Entry and Access Points*

Air intakes constitute the "nose" of the building. Outdoor air dampers are the front end of the intake and signal the location of the intentional air intakes. These are best located one third of the way up the side of the building.

■ ***Many air intakes are located at ground level or below ground level, which makes them particularly vulnerable to possible CB terrorist attacks. Air intakes may also be located on the rooftop — another accessible entry point for CB agents.***

◆ Unintentional Air Intakes — Entry and Access Points

◆ Unintentional air intakes can be located near intentional air intakes or elsewhere in the building. Air is typically drawn into air handling equipment directly from the mechanical room.

◆ Other unintentional air intakes are located at **lower ends** of a building through the **stack effect** or chimney effect.

◆ Apparent roof intakes can be, in reality, sewer vent pipes, cooling towers, kitchen exhausts, lab fume hood exhausts and lavatory exhausts. If terrorists reverse the motors that run these exhausts, the **exhausts will operate as intakes** and may provide an entry port opportunity for CB agents.

◆ Utility pipes or conduits located inside or outside a building can also disseminate CB agents.

■ *The stack effect occurs when heated air (less dense than the colder outside air) escapes from exit points at upper levels of the building. Replacement air enters the building from lower levels (lobbies, elevator, stairwells, entrances, utility chases, laundry and garbage chutes).*

Heating, Ventilation and Air-conditioning Systems (HVAC)

The HVAC system is the predominant pathway and driving force for air movement in buildings. This system includes all heating, cooling and ventilation equipment serving a building. Two of the most common HVAC designs used in modern and public buildings are **constant volume** and **variable air volume**.

◆ Constant volume systems provide a constant airflow to vary the air temperature to meet heating and cooling needs.

◆ Variable air volume (VAV) systems condition the supply air to a constant temperature and ensure thermal comfort by varying the airflow to occupied spaces.

■ *A CB agent released into the ventilation system will find an excellent capture through outdoor air intakes, mechanical room air handling units, and return air intakes. HVAC ductwork and air supply registers provide efficient distribution throughout the building.*

■ *A totally closed system, with very little outdoor air, is also vulnerable to attack. Negative pressure can draw contaminants through unintentional air intakes and indoor air returns. Since there is almost total recirculation of contaminated air, CB concentrations remain high.*

VULNERABILITY ASSESSMENT: BUILDING PROFILE

Three basic types of buildings are most vulnerable to CB attack:

◆ Buildings that house companies or organizations.
◆ Buildings that, if damaged, would disturb the public, the economy, or have the potential to provoke mass hysteria.
◆ Buildings that possess symbolic value.

A building profile describes the features of a building's **structure**, **function** and **occupancy** that impact the security of the indoor environment. The profile provides a baseline understanding of the building's current environmental security, and can help building managers identify vulnerable areas and prioritize budgets for **maintenance** and **future modifications**. Key questions are:

◆ How was the building originally intended to function? Consider building components and furnishings, mechanical equipment (HVAC), occupant population, and associated activities.
◆ Is the building functioning as designed? Determine whether it was commissioned — a process whereby building performance is measured against specifications. Compare the information from the commissioning to its current condition.

Chapter III: Buildings Security

◆ What changes in building layout and use have occurred since the original design and construction?

Developing a Building Profile

A review of construction and operating records, along with a building inspection, can identify areas that require attention to prevent future problems. Baseline data collected for a building profile can facilitate later investigations in the event a problem arises.

◆ **Action**: Collect and review existing records — review design, construction and operating documents.
◆ **Outcome**: Update the description of building system design and operation. Initiate inventory of locations where occupancy, equipment or buildings have changed.
◆ **Action**: Conduct a walkthrough inspection of the building — observe the building and talk with building staff and other occupants. Be alert to problem areas (entry, accessibility and pathways).
◆ **Outcome**: List responsible staff and/or contractors, document evidence of training in securing building systems, and update job descriptions. Record locations that need monitoring or correction.
◆ **Action**: Collect detailed information, including building system condition and operation, CB entry and pathways and occupants.

◆ **Outcome**: Conduct an inventory of system components that need repair, adjustment, or modification/replacement. Complete a plan that indicates airflow directions or pressure differentials in significant areas. Catalog all access and entry points and their locations.

Develop an environmental security plan based on the above outcomes. Review and update that plan annually.

Blueprints and On-site Inspection
Blueprints — two-dimensional drawings that map the location and characteristics of the various building systems' are frequently inaccurate or outdated, due to system additions or modifications.

◆ Reviewing a building's blueprints can help identify exposed areas that are susceptible to attack.
◆ A walkthrough of a building can confirm existing areas and add new areas not present on blueprints.

VULNERABILITY ASSESSMENT TOOLS AND TECHNIQUES

The blueprint review and building walkthrough are qualitative techniques that help assess a building's vulnerability to terrorist attack. Tools to evaluate building systems range from the basic senses to sophisticated and expensive equipment.

Some tools and techniques developed for CB vulnerability assessment include:

◆ Building systems checklist, with a focus on the HVAC/ventilation system.
◆ Micromanometer or equivalent to measure pressure differentials.
◆ Pitot tube to check airflow in ducts.
◆ Water gauge to measure at fans and intakes.
◆ Chemical smoke tubes for observing airflow patterns.
◆ Velometer to assess airflow from diffusers.
◆ Airborne particulate counter.
◆ Organic vapor analyzer.
◆ Odorant tracer.
◆ Tracer decay gas and measurement equipment.

Video and Fiberoptics

Fiberoptic Borescopes
The original military purpose of fiberoptic borescopes was the inspection of rifling for artillery. The basic

system consists of a light source, a fiberoptic image transfer cable, conventional optics, and a barrel to hold the lens and deliver the light source.

These devices provide a very intense light source and an excellent video resolution. Disadvantages include high cost, wires and a 110-volt power source, and unwieldy connections in tight spaces.

Small Diameter Borescopes
Small diameter borescopes (<19mm) can penetrate relatively small openings, but these devices produce images that are fairly dim and small.

Pole-mounted Cameras
This device consists of a camera mounted on a pole with a light source. It is simple in design and operation. The pole-mounted camera is usually a two-person operation, and requires a source of AC power.

Periscopes
A relatively inexpensive option is the mirrored periscope with a flashlight. Primary advantages include relatively low cost, portability and simple design. Negative attributes include poor image quality and illumination.

Fiberscopes
Another useful and readily available remote-viewing device for inspection and clearance is the fiberscope. This is a flexible unit that allows the operator to look

behind walls into equipment chases, crawl spaces, and other areas inaccessible to rigid borescopes.

Remote Video Vehicles

Remote video vehicles can traverse duct-work or other building conduits. They deliver good image quality, while the operator can maneuver them to thoroughly inspect all duct-work.

These devices can also employ grappling arms and tools to extract devices from ducts. Disadvantages include cumbersome size, the need for a large access hole, significant setup and operating time, and cost. Good systems cost in excess of US$50,000, although less expensive — and lower quality — units may cost less than US$10,000.

HARDENING A TARGET

Despite a broad range of potential terrorist targets, not all are equally vulnerable to attack. The building systems most exposed to a CB agent release, for example, are HVAC or ventilation systems. Buildings with air handling units that serve large volumes of building space are especially prone to attack.

By contrast, buildings that utilize numerous small plenum-mounted heat pump units as part of a ventilation network are not likely to face a CB attack.

■ *Not all portions of a ventilation system are likely to be the point of CB agent release. Outside air intakes, return ductwork and the downstream side of the air supply system beyond the coils are all likely points for the introduction of CB agents.*

■ *Place emphasis on protecting the prime CB release locations of the HVAC/ventilation system and the unintentional air intakes in a building.*

■ *Utility pipes may have valves or terminal points accessible at street or below street level. Terrorists can introduce CB agents at these junctions for transport into the building.*

Hardening these potential entry points involves the completion of a vulnerability assessment. Modify, re-engineer and institute engineering and administrative controls to strengthen and defend access points, entry points and pathways throughout the building structure.

Re-engineering Access Points, Entry Points and Pathways

Between varied access and entry points, harden air intakes first.

■ *Many air intakes are located at ground or below ground level. Protect these locations, since they are potential sites of attack.*

An air intake positioned immediately adjacent to pedestrian traffic or parked vehicles presents an opportunity for the terrorist to introduce CB agents into the ventilation system without building entry.

Another site where protection against CB agents is crucial is the **mechanical room**. Most facilities have little or no security in their mechanical rooms. Maintenance personnel frequently leave these rooms unlocked, and doors often have locks that are easily broken. This can create opportunities for terrorists to access such rooms undetected.

■ ***Consider relocating air intakes to inaccessible building areas. Cost, location, aesthetics, building ordinances and practicality will determine whether this is feasible.***

Controlling Pressure Differentials and the Stack Effect

One way to harden a building's ventilation system against CB attack is to bring in more air than is exhausted, creating positive pressure inside the building relative to the outdoors. This minimizes the possible incursion of CB agents through unintentional air intakes.

Protecting Ventilation Systems

◆ High Efficiency Particulate Arrestance (HEPA) filters.
◆ Charcoal filters.
◆ Scrubbers.
◆ Bypass and shutdown mechanisms.
◆ UV lamps.

The addition of 99.99 per cent HEPA filters and activated charcoal filters to the air handling system, in theory, would seem to be a simple fix to trapping CB agents. Yet such filters add significant static pressure to the system, and this necessitates the addition of ventilation fans to overcome the pressure.

■ *Consider adapting technologies used in 'clean' rooms. A small series of panel fans, equipped with HEPA filters on the outlet and an added 30 per cent prefilter, can reside in a 12-inch panel. The fans can activate sequentially as the need for outdoor air increases, and deactivate as the need for outdoor air decreases.*

Another approach is the installation of bypass ductwork to a scrubber, in line with the air stream, followed by an electronic precipitator. The system activates when it detects particulate, gas and vapor in the air handling units. The scrubber itself cleans particulate and gases from the air stream, while the electronic precipitator can remove fine particulate.

The application of ultraviolet lights is a recommended secondary defense, as it successfully controls microbial transmission, contamination and infection in building environments.

SECURING ACCESS AND ENTRY POINTS

Accounting for Occupants
Basic security measures such as ID cards are helpful, but the current state of security technology is electronic, invisible and user-friendly, and offers the ultimate choice in safety and accountability.

■ *Accountability is pivotal in the prevention of a terrorist incident, especially from within a building. Building owners must know who is in their building at all times.*

In most buildings, the main security system is the access control card. Over the years, card systems have become linked to other systems, doubling as an audit trail for both people and equipment.

■ *Proximity cards can record when an individual leaves and enters a facility without the removal of the card from a pocket, wallet or handbag.*

■ *Security personnel can tag equipment, boxes, and other materials, coded to a particular access card. This tag technology is also possible with vehicles.*

Biometric systems identify individuals by reading their eye, palm, voice or thumb signatures. They are advisable for highly secure areas, since they increase security levels by two to three when compared to card systems. Falling prices have moreover made these systems more attractive.

Video surveillance and motion detectors can serve both as deterrents and event recorders.

◆ Closed circuit video imaging devices are getting much smaller and can operate from apertures as small as a pencil eraser.
◆ Smaller sizes yield greater transparency and greater freedom of movement, with image swings of 360 degrees in one second.
◆ Some video systems can transmit as many as 16 different cameras onto one monitoring screen, increasing the effectiveness of security personnel whose task it is to monitor numerous cameras.
◆ New systems allow operators to select which cameras to monitor at particular times of the day and enlarge those images for better viewing.
◆ The use of digital video discs (DVDs) for recording allows for easy imagery searches to quickly identify the exact moment of movement.

■ *Video cameras and signs in plain sight can indicate that an area is under surveillance by both motion and video surveillance. Motion sensors can be part of the macro view of the building air intake areas, and located within the airshaft as well.*

■ *Motion detectors can automatically and simultaneously shut down air handling units, close intake louvers, and sound alarms to security and building operations personnel.*

Chapter IV: Hospitals
Hospital Security Planning
Introduction
Step One: The Planning Process
Step Two: Planning Checklists
Step Three: Response Folders
Step Four: Emergency Tasks
Event Recognition

Hospital Incident Management
Public Safety Response
Planning for Different Types of Attacks

Hospital Security Planning

INTRODUCTION

An act of terrorism has become a growing concern for the public safety community at large. While most health care facilities are essential components in a public safety response system, at present they are poorly prepared to handle either direct or indirect terrorist attacks. The likeliest challenge facing the health care industry may come from an attack elsewhere in the community that leads to the swift arrival of large numbers of contaminated individuals to hospital facilities.

The possibility of a terrorist attack is always present. Hospital administrators, security staff and medical providers need to understand that they will eventually face a criminal or terrorist attack. Such incidents could include **direct events** like an assault against a facility or **indirect attacks** that may necessitate the provision of care to large numbers of contaminated patients.

Facility planning should rest on the adage that the weakest link in the planning process is typically an underestimation of the ingenuity and resourcefulness of criminals or terrorists, who are determined to fulfill their goals.

STEP ONE: THE PLANNING PROCESS

The goal of a "living" planning guide is to provide the planner with the information and necessary tools to plan, prepare, respond to, and then integrate planning with all public safety organizations that will play a role in the event of an attack on a facility (Table I).

A measured plan and effective coordination with public safety organizations will further joint agency integration and the proper utilization of assets.

There are a few items to consider first when planners begin to develop a plan:

1. Is there a need for an attack plan?
2. Who has the ability to help develop a plan?
3. With what public safety organizations should planners coordinate or make contact?
4. How should the planning process begin?

Q1. Do I really need to plan for an attack?
A1. The answer is an unequivocal **yes**. Terrorists and criminals continually develop new ways to sow fear, panic and terror in the minds of civilians. In recent years, the typical targeting methodology favored by terrorists has shifted from **standard** to more **ad hoc** or outré methodology. World borders are increasingly porous, and terrorists will continue to strike at locations that authorities once considered safe havens, or immune from attack.

Table I: Players in a Terrorist Incident			
Organization	**Local Public Safety**	**State Public Safety**	**US Federal Agencies**
Security director	Police	State police	FBI
Security staff	Fire/EMS	State fire marshal	FEMA
Hospital administrator	Local emergency management	State emergency management	EPA
Unit managers	Local public health	State medical examiner	DoD
	District attorney	State public health	DoJ
		State attorney general's office	DoT
			DHHS
			DoE
			ATF
			US Treasury

Q2. Who has the ability to help develop a plan?

A2. **Local, state and federal agencies** all have departments within their organizations with the ability to assist the public conduct emergency planning and response.

Q3. How will I coordinate or make contact with the necessary public safety organizations?

A3. One easy way to begin the planning process is to contact a local or state emergency management office. These departments and agencies have the ability to reach out to other members of the public safety community to coordinate a well-rounded meeting. **Offices of Emergency Management (OEMs)** are **one-stop shopping** departments for public safety officials.

Q4. How should I begin planning?

A4. It is important to begin the planning process by first establishing the purpose and objective of the plan. Develop a one-paragraph statement that describes the plan, which will provide a framework for all participating planners.

STEP TWO: PLANNING CHECKLISTS

1. Developing an Emergency Operations Plan (EOP).

It will become clear in the early planning stages that most **Emergency Operations Plans (EOPs)** have a similar layout. Typical plans assign responsibility to an organization or to individuals, in order to carry out specific actions at projected times and places during emergencies. **An EOP sets organizational relationships and authority guidelines**. It also details how responders will protect people and property in a given situation. An EOP will describe how facility personnel will integrate and utilize equipment, personnel, resources and supplies during an event. It will establish authorities for particular functions during an emergency. Finally a good EOP will address **mitigation strategies** following an event.

2. Conduct a Vulnerability Assessment of the Facility

Derive a list of vulnerabilities from local resources, or from the following chart (Table II). Seek the advice and assistance of the planning team to ensure the development of the most comprehensive list possible, to better prepare for an attack.

Table II: Sample Vulnerability Assessment

❏ 1. Minimal security.
❏ 2. Easy access to the public.
❏ 3. Mixed business interaction with minimal contact. An example of this might be a clinic or office that operates in a location away from the secure hospital and has no reception point. This provides a terrorist the ability to conduct surveillance and gain intelligence without being challenged.
❏ 4. Central receiving. Personnel assigned to central receiving may be less likely to question suspicious packages because they are unfamiliar with the size and types of packages the organization receives.
❏ 5. Multilevel buildings. Facilities with many levels present a challenge to all planning documents. First, they are difficult to evacuate in any situation. Second, they provide an opportunity for a terrorist to hide devices or themselves. Third, elevator shafts and stairwells can carry contaminates through a building while individuals attempt to evacuate.
❏ 6. Groups in an area.
❏ 7. Parking garage.
❏ 8. Restricted exits/fire escapes.
❏ 9. Inadequate air handling systems.
❏ 10. No control of HVAC or other utilities in the building.
❏ 11. On site storage of HazMat and Bio-waste material.
❏ 12. Close proximity to other industrial businesses.
❏ 13. Routes of passage for hazardous material in close proximity. Does a major trucking route pass directly in front of the location?
❏ 14. Easy escape route via major roadway.
❏ 15. Concealment locations in close proximity. This could allow an attacker to hide his/her location in an attack and to execute a secondary attack.

Table II: (continued)

❏	16. No visible security presence.
❏	17. Routine response to unscheduled events. Do personnel take events for granted? Do they respond to "bomb hoaxes" with reluctance?
❏	18. Do personnel challenge unscheduled or unannounced visitors?
❏	19. Has there already been an attack or incident of some kind?
❏	20. Does the organization sponsor or support an event that may be controversial?

After compiling a vulnerability list, write the questions in one column (Table III). The planning team should conduct an assessment using the vulnerability list. Place a check in the column applicable to the target. Total check marks in each of the columns. The higher the column total, the more susceptible is the target to a terrorist attack.

Following a determination that there is a need for increased security at the facility, approach the planning process aggressively.

STEP THREE: RESPONSE FOLDERS

Planners can now undertake a methodical and systematic step-by-step planning process for a facility. Clearly there is a need to develop a planning document. This exercise will provide guidelines to help planners build and plan a **response folder** for a given location. This document will provide planners and response agencies with the necessary information to respond to an incident at a facility.

Table III: Sample Planning Checklist		
Location Name (eg Main Building)	Location Name (eg Parking Garage)	
✓	❏	Minimal security.
❏	✓	Easy access to the location.
❏	❏	Central receiving.
✓	❏	Multilevel buildings.
❏	✓	Groups in an area.
❏	✓	Parking garage.
❏	✓	Restricted exits/fire escapes.
❏	✓	Inadequate air handling systems.
❏	✓	No control of HVAC or other utilities in the building.

		Table III: (continued)
❏	✓	On site storage of HazMat and Bio-waste material.
❏	❏	Close proximity to other industrial businesses.
✓	❏	Routes of passage for hazardous material in close proximity.
✓	❏	Easy escape route via major roadway.
✓	❏	Concealment locations in close proximity.
❏	✓	No visible security presence.
❏	✓	Routine response to unscheduled events.
❏	✓	Do personnel challenge unscheduled or unannounced visitors?
✓	❏	Has there been an attack or incident of some kind?
✓	❏	Does the organization sponsor or support event (s) that may be controversial?
7	10	Total

RESPONSE FOLDER CONTENTS:

SECTION ONE

1. General site information:
   a. Role of the facility in the community.
   b. Economic impact on the local area.
   c. Employment.

2. Location:
   a. Detailed strip maps to the location from the nearest airports, interstate highways and other principal routes.
   b. Surrounding area or local area maps (counties, cities and towns within a 30-mile radius).
   c. Maps to additional facilities or support facilities.

3. Location boundaries:
   a. Area.
   b. General data.

4. Facility description:
   a. Photos.
   b. GPS coordinates.
   c. Map references.
   d. Azimuth and distances to local features.
   e. Grounds description.
   f. Helicopter landing zones with approach obstacles (trees, lights, radio towers and wires).
   g. Facility construction.
      i. Roof entrances.
      ii. Roof type and strength.
      iii. Fire escapes and ladders.
      iv. Stairways.
      v. Elevators.
      vi. Loading docks.

5. Security:
   a. Protective measures.
      i. General defensive strategies.
      ii. Protection strategy.
      iii. Security force in facility.
      iv. Security communications.
   b. Restricted areas.
      i. Reason for restriction.
      ii. Security measures in restricted areas.
   c. Security systems.
      i. Intrusion detection.
      ii. Surveillance measures.
      iii. Fire monitoring system.

6. Electrical power sources:
   a. Security system power supply.
   b. Backup power.
   c. Operations power.
   d. Diagrams (panel locations).
   e. Matrix.
      i. Area/equipment affected.
      ii. Breaker.
      iii. Panel.
      iv. Panel locations.
   f. Impact of loss of electrical power.

7. Ventilation systems with photos:
   a. Air-conditioning unit.
   b. Size.

   c. Alarms.
   d. Monitoring.

8. Telephone system:
   a. Panel locations.
   b. Phone numbers to outside lines.
   c. Servicing company and point of contact.
   d. Cell phone availability and location.

9. Medical support:
   a. Location.
   b. Capabilities.

10. Fire protection:
   a. On-site firefighting equipment.
   b. Sprinkler system.
   c. Extinguisher locations.
   d. SCBA capabilities and locations.
   e. Hydrant locations.
   f. Halon capabilities.
   g. Mutual aid.
   h. First-in unit response times.
   i. Established command cost locations.

11. Hazards:
   a. Toxic chemicals.
   b. Biological waste.
   c. Infectious waste.

12. Water:
    a. Potable water location.
    b. Water shut-off locations.
    c. Emergency storage capabilities and storage locations.
    d. Maintenance company.

13. Sewer:
    a. Manhole locations.
    b. Routing diagrams and dimensions.

14. Vehicles:
    a. Types and number usually found.
    b. Locations of parking lots.
    c. Diagrams and capacities of enclosed or underground parking facilities.

15. Staffing:
    a. Key personnel data.

SECTION TWO

1. Building layout — floor diagrams:
    a. Doors.
    b. Interior windows.
    c. Rooms.
    d. CCTV locations.
    e. Other pertinent security information.

2 Operational considerations:
   a. Likely hostage-holding areas.
   b. Potential adversary holding areas.
       i.   Fields of fire.
       ii. Concealment of additional weapons or devices.

3. Critical paths:
   a. Critical path diagrams (with photos).
   b. Critical path doors.
       i.   List of doors along path with descriptions.
       ii. Door information matrix.
            1. Interior/exterior.
            2. Glass size (wire).
            3. Width.
            4. Height.
            5. Thickness.
            6. Lock type.
            7. Door opening direction.
            8. Material.
            9. Keys and key control.

In addition to the creation of a **response folder**, it is also important to consider the placement of the initial command post. **In any terrorism incident, the first few minutes will be chaotic.** To minimize the extent of this disorder, establish **primary** and **secondary** locations beforehand, where security staff and internal emergency response personnel will regroup in the event of an incident. When planning the location of

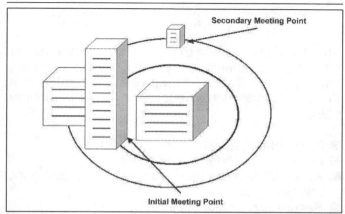

Figure 1

this rendezvous point, consider that terrorists may target this location as a secondary objective. Locate the rendezvous location within two concentric circles (Figure 1). This methodology will give responders and planners the ability to decrease their profile and minimize the risk of first responder injury from terrorist secondary devices.

IMPORTANT NOTE
Obviously, a facility's response folder is a valuable document for a terrorist or criminal planning an attack. Therefore, planners **must** keep this document under strict control, and make it available **only** to those individuals who have a required security need to review or examine the material within the folder.

STEP FOUR: EMERGENCY TASKS

After they establish an **EOP** and a **response folder**, planners should establish priorities in the event of a terrorist incident. The following list is a starting point to help planners achieve some or all of the required objectives:

■ Notify emergency response agencies.
■ Notify key hospital planning staff.
■ Activate internal emergency response personnel and resources.
■ Secure and isolate the incident area.
■ Account for the safety of the internal and external population (external refers to people or personnel immediately outside of the facility).
■ Establish the Incident Command Post (ICP).
■ Assess the situation.
■ Begin evacuation or sheltering in place, if required.
■ Provide updates to people located throughout the facility (rumor control).
■ Recommend and coordinate personnel protection measures.
■ Provide overall coordination of the incident.
■ Establish a system to handle public inquiries.

ZONE CONTROL

Regardless of whether the type of terrorist attack is relatively normal (**armed attack, bombing**), or an unusual occurrence (**WMD attack**), planners and responders must establish initial areas for continued safe operation. Planners must consider both short-term response operations and extended operations that could last hours or days.

An especially difficult planning concern is the fact that a hospital facility may receive patients from another developing situation, or need to begin to evacuate patients. While planning for zones of operation always establish:

1. Safe ingress to the CP and staging locations.
2. Safe egress of emergency vehicles.
3. Control of ingress and egress around the facility staging areas.
4. Evacuation areas and routes.
5. Inner and outer perimeters.
6. Media holding areas.
7. Hot, Warm and Cold Zones to limit contamination.
8. Decontamination locations.

IMMEDIATE ACTIONS

When first arriving on-scene following a potential incident, the **correct identification** of the nature of the incident is the first step in first response. There may be an initial period of **chaos**. This will generally begin following an incident, when most people are in some form of **shock**. It is during this time that responders and planners must quickly take control. Below are guidelines that responders and planners can use to **Gain Control Of The Situation (GCOTS)**.

Gain Control Of The Situation:

1. Size-up the situation.
2. Evaluate the situation.
3. Set incident priorities.
4. Determine the potential for further harm, injury or destruction.
5. Establish one incident command post.

Once planners and responders GCOTS, they must employ additional methods as the incident unfolds. However, if chaos increases rather than decreases over time, and responders lose control, they should fall back on GCOTS procedures to reorient themselves and others on priority concerns.

In addition to GCOTS, responders will need to perform other Immediate Actions (IAs). Generally these IAs will focus on life safety: **this is always the first priority of any EOP or response plan**.

IAs:

1. Remove any continued threat to life safety.
2. Evacuate the area.
3. Begin fire suppression, if able.
4. Begin emergency medical procedures.

By performing these IAs, responders will have achieved the majority of the primary objectives of most emergency response agencies.

Because of the complexity of a terrorist event, security planners should begin by delineating priorities that guide the preparedness process. Coordinate these objectives with other members of the emergency response community. Planners should rank the priorities in this order:

1. Protection of current patients, staff and the facility.
2. Provision of the best possible medical care to patients who arrive at the institution for treatment.
3. Environmental protection, external to the facility, in a large-scale event.

Planners can make certain assumptions in order to simplify planning. **The exposure site will probably be remote from the hospital** (for example, the hospital will not be within the primary release area of an incident, but will still receive patients injured in the incident). If the situation is otherwise, and the hospital **is** within the primary release area, this may indicate that facility evacuation or "sheltering in place" is necessary.

EVENT RECOGNITION

In any suspected terrorist event, it is essential that response personnel identify terrorist victims before they enter the medical facility. Security personnel must receive training in early recognition techniques, and planners should plan to station them at hospital entrances. Security personnel should **immediately** notify management personnel when they suspect a problem. They should be prepared to protect themselves, and don **personnel protective equipment (PPE)** or close the facility to prevent possible contamination until the establishment of proper response procedures.

Even so, it is reasonable to expect that some contaminated individuals may gain entrance to the facility. Responders should handle such situations on a case-by-case basis and employ a rational approach.

Hospital Incident Management

Through the use of the **Incident Management System (IMS)** planners and responders will have the ability to effectively manage an incident prior to the arrival of outside resources. If responders have a system in place, designed to address all the critical functions of a facility, along with a standardized response procedure, they will more efficiently be able to coordinate the outcome of the event.

Hospital management and planners should incorporate the principles of the ICS into the hospital's emergency preparedness plan. The use of this system will enable **Health Care Facility (HCF)** staff to fully integrate their activities with community emergency response assets. Although initial response efforts will center on **decontamination** and **treatment areas**, other hospital departments will play vital roles. For instance, security officers will need to direct the flow of casualties and vehicles to prevent facility compromise, and prevent unauthorized access to the decontamination and treatment areas.

A terrorist attack with the possibility of a widespread **chemical** or **biological** incident will result in extended operational periods. Management and planners should coordinate activities to relieve staff from physically taxing activities such as **patient decontamination**.

Once an incident occurs, the organization should follow **Standard Operating Procedures (SOPs)** contained within its EOP. After this step, the management process is simple and straightforward. It is important that initial responders take charge after they receive notification of an incident:

1. Follow guidelines established in GCOTS.
   a. Conduct a situation size-up.
   b. Evaluate the situation.
   c. Set incident priorities.
   d. Determine the potential for further harm, injury or destruction.
   e. Establish one incident command post.

2. Activate a common communications method.
   a. Internal communications.
   b. Radio (not in case of bomb threat).
   c. PA system.
   d. Cell phone (not in case of bomb threat).
   e. Interactive paging system (not in case of a bomb threat).
   f. E-mail.

3. Begin a process to:
   a. Gather information: The initial first responder to a situation within the facility will need to gather as much information as possible, through observation. Given that the situation could be the result of a terrorist attack, responders should observe the situation up-wind of the scene and away from any continued, perceived hazard (Figure 2). The

responder's level of education and experience will help to evaluate and process the information as the responder views it. The responder must always remember that even though a hazard may not be visible to the naked eye, it may still exist.

b. Key points to note during the information-gathering process:

    i. Unusual signs and symptoms.
    ii. Presence of dead animals or people.
    iii. Unattended items such as sprayers or packages.
    iv. Unexplained odors.
    v. Anything out of the ordinary.
    vi. Environmental information.
    vii. Time of day.
    viii. Current weather.
    ix. Wind direction.
    x. Lay of the land.

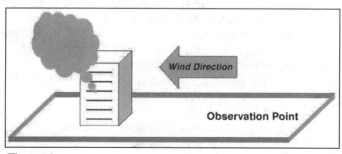

Figure 2

Chapter IV: Hospitals

■ *Regardless of the situation it is important that responders carry out IAs, quickly gather all information, and continually reassess this information as the situation develops.*

   c. Estimate further damage or potential harm. Use previously gathered information to make a series of predictions to assess the potential for further injury. This estimate will include damage assessment, hazard identification, vulnerability and risk determination.

     i. Damage assessment involves a determination of the **amount of damage** that has already occurred.

     ii. Hazard identification involves a determination of **what** has caused the incident to occur, **where** it is, **what** it can do, and **how much** is there.

     iii. A vulnerability assessment involves a determination of **who** and **what** is at further risk in and around the location.

     iv. Determining risk is a procedure to ascertain if the situation will get worse.

   d. Establish goals and priorities.

     i. When establishing response goals, make these short and to the point.

       1. Life safety.

       2. Protection of critical systems.

       3. Incident stabilization.

e. Evaluation. The evaluation process helps to determine if responders are properly executing the IAP, which planners and public safety response agencies will have already established. Evaluation can also help identify possible errors and unforeseen issues during the initial planning phase.

Public Safety Response

In the first few minutes of an incident, local police, fire or EMS agencies may receive the initial notification. The notification may also come from planners or management, who are initiating an internal EOP.

Planners and management who follow the steps laid out in the initial sections of this chapter will have already contacted local authorities as well as a host of others responsible for public safety. If others from the public safety community helped to develop the facility's planning process, they will already have a working knowledge of facility operations and current capabilities during an incident. **They will expect to receive a copy of the response folder upon arrival**.

During the initial phases of a response to an incident there are many issues that will arise which may create more chaos and confusion on-scene. **The individual responsible for the facility must assume responsibility to ensure the proper transmission of information to arriving public safety units.** Issues to consider during the initial response to an incident include:

1. First response units may be unfamiliar with the facility, its planners and its plan.

2. A level of apprehension may be present on the side of the first responders, due to the nature of an attack or incident, such as:
   a. Armed attack.
   b. Hostage situation.
   c. Product tampering.
   d. Active explosive device.

Armed Assaults

Terrorists stage armed assaults to strike selected targets quickly. This form of attack comprises few operational aspects, and requires limited planning and resources. One type of armed assault is the stand-off type of attack, using mortars, rockets or remotely detonated explosive devices. Terrorists who employ stand-off attacks can strike their intended targets but remain safe from nearby security forces.

Another form of armed assault is the close-range strike. Executed very rapidly and effectively, close-range strikes are comparable to drive-by shooting techniques popular with Los Angeles street gangs, for example.

Hostage-Taking

Terrorists can seize one or several individuals, whether to secure the release of their imprisoned colleagues, gain concessions, sue for ransom, or attract publicity.

Terrorists usually carry out extensive planning when they elect to take hostages, including long-term prior surveillance and reconnaissance. Typically, only **domestic** terrorist organizations take hostages, within the local environment. Most terrorists who seize hostages also tend to be advocates of political, religious or ethnic causes.

Product Tampering

Terrorists can employ product tampering to extort funds, inflict large-scale monetary damages, or embarrass a government. Terrorists have the capacity to poison food or pharmaceutical goods. The threat of such an activity alone can cause a targeted organization to discontinue a product or service, and this can lead to a loss of revenue or loss of confidence in a government.

Bombings

Finally, bombings are a long-time standard tactic favored by terrorists. For the terrorist, bombings are attractive because they provide immediate casualties and news coverage, leading to immediate attention to the terrorist's cause or political opinion.

■ *Bombings are tremendous media events, and the larger the explosion, the larger will be the extent of media coverage.*

Bombs are some of the most effective instruments with which to launch a terrorist attack. Explosives and technology are easily accessible, reasonably affordable, and often require minimal technical expertise. Bombing operations are low-risk, since terrorists can trigger such incidents **clandestinely**. Sophisticated timing devices also make the handling of bomb materials fairly safe.

Managing First Responders at Bombing Incidents

Planners and managers need to recognize that bomb incidents trigger uneasiness in responders. This tension renders it doubly important that management provide responders with all the information detailed in preceding sections, quickly and efficiently.

While managers establish a command post (as described earlier), first responders will be doing the same. If the first responder units do not co-locate with management, then management should contact first responders and provide them with the information that planners collated earlier. **It is important that management and planners do not relocate from their established location unless proper authorities otherwise direct them to do so**.

■ *Management must endeavor to maintain continuity of operations. Hospital units and departments will be attempting to contact management at the predetermined command post location.*

Depending upon the incident, the public safety response may be limited or it may be massive in scale. It is important that management **maintain a sense of control and reason** over the course of the first response. The arriving first responders and additional assets may determine that they have reasons to request additional assistance, depending upon their capabilities and the type of incident. **Remain accommodating and helpful**. This approach will ensure a coordinated and goal-oriented response.

PLANNING FOR DIFFERENT TYPES OF ATTACKS

Planning for terrorist attacks should always be all-inclusive. Never be *completely* unprepared for a given attack type. However, instead of planning equally for all types of attack scenarios, **allocate time and resources based on which attacks are most likely**. Include the possibility of others, but stress the likeliest attack types.

For example, in a hospital, there is a **high possibility** that an injured individual might open fire on staff with a **hand-held weapon**, and then take emergency room hostages. The possibility is **low** that a terrorist would utilize a **nuclear weapon** in a hospital.

As a corollary to this scenario, however, there is a **high probability** that if a terrorist group detonated a **nuclear or radiological** device in a major city, then a hospital would be responsible for treating victims of the blast and radiation. Planners and management should gauge the **probabilities** of certain terrorist attack scenarios, and focus on the most probable cases, but nonetheless give some attention to low probability incidents.

Chapter V: Educational Institutions

Educational Institutions

Introduction

Definitions

It Could Happen! What Do We Do First?

Vulnerability and Hazard Analysis

Action Planning Checklist

Responding to Crisis

Direction and Control

Command Post

Overall School Responsibility

Before Help Arrives

Emergency Procedures

Direction and Control

Responding in the Aftermath of Crisis

Crisis Procedure Checklist

Educational Institutions

Introduction

Unanticipated tragic events can quickly escalate into a school-wide catastrophe if the administration does not deal with events immediately and effectively. Knowing what to do when a crisis occurs can minimize chaos, rumors and the psychological impact of an event on students and the community.

When a disaster strikes, teachers and school staff members will be torn between their own reactions and the need to deal with **student** reactions. At this time, the **faculty** and the **administration** are likely to be least prepared to think quickly. Advance planning can ensure this process runs far more smoothly than when a tragedy strikes a school that has no pre-formulated plan.

Definitions

Crisis: *This is a sudden, generally unanticipated event that profoundly and negatively affects a significant segment of the school population and often involves serious injury or death. The incident will affect a large number of students and staff. The psychological and emotional impact will be moderate to severe, and outside assistance will be necessary.*

Crisis Team: *This is a group that consists of administrators, school psychologists, counselors and other persons who are responsible for handling media, traffic, logistics and information. It is the decision of the principal when or whether to activate the Crisis Team.*

Auxiliary Team: *This pre-established and trained group includes representatives from law enforcement, mental health agencies, the medical community, clergy, parents, patrons and school personnel from other districts.*

Calling Tree: *This mechanism allows the team to receive immediate notification. The list of numbers should reside in the principal's office. The secretary to the principal should provide training to other secretaries to ensure planning for all necessary communications. This should include a plan for internal communications between staff members.*

Crisis Kit: *Establish in each principal's office a container with name tags, notebooks, pens, markers, hand radios, batteries, first aid supplies and tape. Also prepare separate placards labeled with simple group names or directions, such as:*

1. PARENTS
2. COUNSELORS
3. MEDIA
4. CLERGY

5. VOLUNTEERS
6. KEEP OUT

Caution:
Tape and other supplies should also be in the crisis kit. Have copies of student records, especially health and identification papers, ready to send quickly to the hospital in the event of injuries. Also send the hospital a current yearbook. A laptop computer, printer and a copier must be available for immediate use.

It Could Happen! What Do We Do First?

Priority 1: Protect Students

a. Get students out of harm's way.
b. If a sniper is outside, use care in getting students out of the line of fire.
c. Keep covered and keep students covered until all is clear.
d. If a hostage situation exists, perform a lockdown.

1. Have someone call 911 and continue to provide first aid to victims.
2. Have staff ready the Crisis Kit, and notify the principal.
3. Continue to provide first aid to victims until EMS arrives on-scene. If possible, identify incapacitated students by using nametags or markers on their wrists or ankles. Be careful if at all possible not to alter the surroundings, since authorities will later investigate the crime scene. Clear uninjured students from the immediate area, and direct them to a pre-designated safe gathering area.
4. Instruct a secretary to activate the calling tree, before telephone lines become overloaded.
5. The principal may need to send the rest of the students to the pre-designated location. It may be back to certain **classrooms**, or the **cafeteria**; the principal should make this decision

based on the situation. The principal needs to notify and assure all individuals in the building, by way of a bell or some other measure, that all is safe. Teachers and staff who are not tending to victims need to be with, and give support to, the rest of the students.

■ *One designated staff person needs to be at the hospital to help with identification, to support parents, and take the information notebook from the Crisis Kit.*

6. Set up tables and placards to give information and directions.
7. Set up pre-designated rooms for media, family, etc.

VULNERABILITY AND HAZARD ANALYSIS

Hazards include any threat with the *potential* to disrupt the school, cause damage and create casualties. Examples of **natural hazards** include severe storms, earthquakes and fire. Examples of **human-caused** or **technological** hazards include hazardous material incidents, nuclear power plant accidents, dam failure, terrorism and/or civil unrest.

A vulnerability and hazard analysis is an important first step in the process of assembling an EOP. The analysis will identify and set procedures, training requirements and the resources needed to respond to, and successfully recover from, an incident that results from any one of the aforementioned hazards.

In addition to the identification of hazards that may impact the school, address:

1. **Probability** – How likely is it to occur?
2. **Vulnerability** – If it occurs, what effect will it have on the school, its students and staff?
3. **Frequency** – Historically, how often have certain hazards occurred, such as earthquakes, bomb threats, armed attacks, etc?
4. **Speed of onset** – How much warning is there likely to be?
5. **Duration** – How long will the hazard likely last?

How to Conduct a Hazard Analysis

Although a brainstorming session that includes staff members is a good way to address possible hazards and raise faculty consciousness, much of the work is already complete. The administration simply has to contact one or more **public safety organizations**, and then apply this information to a specific location and school. (For example, public safety organizations will have identified and possess information about hazardous chemicals that governments or companies store or transport near the educational institution). What follows are valuable techniques towards the realization of a hazard analysis:

Preliminary Steps to Mitigating Potential Hazards and Vulnerabilities

Before an incident occurs, one of the best methods to prevent an act from occurring at a school is to involve school officials in the design of new school facilities or remodeling of existing property. Keep these steps in mind:

1. The administration and planners should insist on their involvement in the **design** process. They should work with architects and construction personnel in the early stages of **remodeling** to provide input on how school design techniques can improve supervision and safety procedures.
2. Carefully consider the **location** of common areas. These areas comprise sites frequently used for

after-hours events (e.g. gyms, auditoriums, cafeterias and libraries), and other key locations. Take steps to control access, and limit the use of these areas. **Restrict access to all areas of the school in the evening.**

3. Review parking lot placement, size and related factors. Work to best facilitate safe movement and supervision.

4. Consider the importance of **line of sight** in hallways and areas under supervision.

5. Take into consideration opportunities for natural surveillance and supervision. Locate areas of greater activity or higher risk where there will be higher levels of adult supervision.

6. Involve school security officials, school resource officers, and/or outside school safety specialists in the planning and design of new or remodeled facilities.

ACTION PLANNING CHECKLIST

Prevention-Intervention-Crisis Response

1. *What To Look For — Key Characteristics of Responsive and Safe Schools*

Does the school have characteristics that:
❏ Are responsive to all children?

2. *What To Look For — Early Warning Signs of Violence*

Has the school taken steps to ensure that all staff, students and families:
❏ Understand the principles underlying the identification of early warning signs?
   Know how to identify and respond to imminent warning signs?
❏ Are able to identify early warning signs?

3. What To Do — Intervention: Getting Help for Troubled Children

Does the school:
❏ Understand the principles underlying intervention?
❏ Make early intervention available for students at risk of behavioral problems?
❏ Provide individualized interventions for students with behavioral problems?

❑ Have school-wide preventive strategies in place
   that support early intervention?

4. What To Do — Crisis Response

Does the school:
❑ Understand the principles underlying crisis
   response?
❑ Have procedures to intervene during a crisis to
   ensure safety?
❑ Know how to respond in the aftermath of tragedy?

RESPONDING TO CRISIS

Violence can happen at any time and anywhere. Effective and safe schools, however, are well prepared for any potential crisis or violent action.

Accordingly, crisis response is an important component of a violence prevention and response plan. Two components that need to be addressed in that plan are:

♦ Intervention during a crisis to ensure safety.
♦ Response in the aftermath of tragedy.

In addition to the development of a contingency plan, schools should adequately prepare their core violence prevention and response team. The team not only **sets plans** for what to do when violence strikes, but also **ensures that staff and students know how to behave in such situations**. Students and staff will feel secure if there is a well-conceived plan, and if everyone understands what to do or whom to ask for instructions.

Principles of Crisis Response
As with other interventions, crisis intervention planning is built on a foundation that is safe and responsive to children. Crisis planning should include:

♦ Training for teachers and staff in a range of skills, including procedures to better deal with escalating classroom situations and response to a crisis.

♦ Reference to district or state procedures. Many
  states have recommended crisis intervention manuals that are available to their local education
  agencies and schools.
♦ Involvement of community agencies, including police, fire and rescue, as well as hospital; health;
  social welfare; and mental health services.
♦ Provision for the core team to meet regularly to
  identify potentially troubled or violent students, or
  other situations that may become dangerous.

Effective school communities also make a point to
find out about federal, state and local resources
that are available to help during and after a crisis, and
to secure their support and involvement before a
crisis.

Actions During a Crisis To Ensure Safety

Weapons used in or around schools, fights, bomb
threats or explosions, as well as natural disasters,
accidents and suicides call for **immediate and
planned action**, followed by long-term, post-crisis
intervention. Planning for such contingencies
reduces chaos and trauma. The crisis response
part of the plan also must include contingency provisions. Such provisions may include:

♦ Evacuation procedures and other procedures to
  protect students and staff from harm. It is critical
  that schools **identify safe areas** where students

and staff should go in a crisis. It is also important that schools **practice evacuation** of staff and students from the premises in an orderly manner.

♦ An effective, foolproof communication system. Individuals must have **designated roles and responsibilities** to prevent confusion.

♦ A process for securing **immediate external support** from law enforcement officials and other relevant community agencies.

The core team should regularly **monitor** and **review** all provisions and procedures.

Just as staff should understand and practice fire drill procedures routinely, they should also practice response to the presence of firearms and other weapons, severe threats of violence, hostage situations and other acts of terror. School communities can ensure that staff and students practice such procedures in the following ways:

♦ Provide in-service training for all faculty and staff to explain both the plan and what to do in a crisis. Where appropriate, include community police, youth workers and other community members.

♦ Produce a written manual, a small pamphlet or flip chart to remind teachers and staff of their duties.

♦ Practice response to imminent warning signs of violence. Make sure all adults in the building have an understanding of **what they can do to prevent violence** (e.g., vigilance, knowledge about when to find help, good problem solving, effective anger management, and/or strong conflict resolution skills) and how they can **safely support each other**.

DIRECTION AND CONTROL

The following are examples of resources that are useful in a response to an incident in a school facility.

Incident Management Team
♦ Emergency plan and procedures.
♦ Roster of school staff.
♦ Contact list of students' families.
♦ Assignment list of emergency roles and responsibilities.
♦ School diagram (exits, shelters, fire extinguishers, emergency supplies, etc.)
♦ Radios and pagers.
♦ Cellular phones.
♦ Bullhorn.
♦ Flashlights and batteries.
♦ Radio with extra batteries.

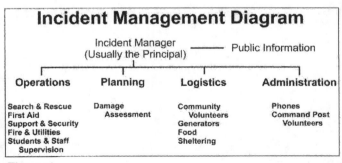

Figure 1

♦ Clipboards, paper, pens and pencils.
♦ Inventory of food and medical supplies.

(See **Figure 1** for an incident management diagram of crisis roles and responsibilities.)

COMMAND POST

Establish a space for the Incident Management Team to carry out its job. Ideally, the room should be somewhat removed from the confusion in the main school area. There should be tables and chairs present for team members. An **alternate area** is also necessary, in the event that the primary location is either unavailable or inaccessible. At a minimum, the room or space should include the following items on this checklist.

■ *For an outdoors command post, determine what items on this checklist can be useful outdoors. Be sure to also develop a list for outdoor use only.*

❑ Flip charts with marking pens, masking tape, thumb tacks/push pins or dry ink board with special pens.
❑ Detailed street maps and freeway maps of the area.
❑ Job action packets that contain assignments and checklists.
❑ Badges and/or colorful vests.
❑ Rotary file with telephone numbers and addresses of local vendors, pharmacies, contractors, etc.
❑ Flashlights/light sticks and extra batteries.
❑ Portable radio.

❏ Walkie-talkies (at least two sets) to use inside the building.
❏ Portable/cellular telephones.
❏ Message pads, pens and pencils.
❏ School food/supplies inventory.
❏ Drinking water, snacks and food for the ICS team.
❏ Extra first aid kit(s).
❏ First aid book.
❏ FAX machine.

The following supplies, classified by the team that will use them, can reside in a box in the **command post** so that they are readily available. If materials are in proper storage, the administration can use the room for other purposes in non-emergencies, then easily convert the room into a command post when necessary.

First Aid Team
♦ Health information on students and staff.
♦ First aid and medical supplies.
♦ Blankets and inflatable mattresses.
♦ Flashlights and batteries.
♦ Radios and pagers.
♦ Clipboards, paper, pens and pencils.
♦ Status tags.

Search and Rescue Team
♦ Roster of employees and students.
♦ School diagram and search assignments.

- ♦ Flashlights and batteries.
- ♦ Radios and pagers.
- ♦ Axes, crowbars, shovels and ropes.
- ♦ Master keys and bolt cutters.

Fire/Utilities Team
- ♦ Fire extinguishers.
- ♦ Shovels and axes.
- ♦ Gloves.
- ♦ Radios and pagers.
- ♦ Tools to turn off utilities.
- ♦ School diagram (exits and utility turn-offs).

Support and Security Team
- ♦ Radios and pagers.
- ♦ Cellular phone(s).
- ♦ Signs to post.
- ♦ Clipboards, paper, pens and markers.
- ♦ Master keys.
- ♦ Shelter plan and procedures.
- ♦ Diagram of the school.
- ♦ Roster of school students and staff.
- ♦ Battery-operated radio.
- ♦ Radio and pagers.
- ♦ Cellular phone(s).
- ♦ Flashlights and batteries.

Evacuation Team
- ♦ Optional team that can be moved to incident management.

♦ Evacuation plan and procedures.
♦ Diagram of school.
♦ Roster of school students and staff.
♦ Radio and pagers.
♦ Cellular phone.
♦ Signs to post and markers.
♦ Master keys.

Damage Assessment Team
♦ Gloves.
♦ Heavy shoes.
♦ Flashlight.
♦ Hazard tape.

■ *Develop a Checklist to Ensure Response and Readiness*

This checklist highlights activities under each response function that planners need to address in order to ensure an effective response to an event.

OVERALL SCHOOL RESPONSIBILITY

School Principal and/or Staff/Parent Planning Committee

❏ Maintain staff awareness of disaster threats.

❏ Hold drills and arrange or conduct training.

❏ Inventory staff members to determine skills that may be useful in disaster planning: e.g. first aid skills, CPR certification, bilingualism, ham radio operation ability.

❏ Make sure the command post area contains a floor plan of the school, a current personnel roster, critical telephone numbers and a dependable communications system.

❏ Designate a media spokesperson.

❏ Develop a release plan for employees. It should take into account family and other responsibilities outside the workplace.

❏ Promote employee family preparedness.

❏ Promote student/family preparedness.

❏ Encourage staff and students to keep an emergency kit (food, water, flashlight, medication and sturdy shoes) in a safe, accessible place.

Command Post

❏ Assemble all necessary information and supplies/material (emergency plan, situation board, maps, markers, radios, walkie-talkies and personnel rosters) at the designated command post location.

❏ Define and assign functional responsibilities (incoming reports, display, response decisions and communications) to staff members as specified in the emergency plan.
❏ Identify and train all staff.
❏ Participate in all planned drills and exercises, and practice activation of the command post.

Support and Security
❏ Develop a plan to control access to the school and to record people who leave and arrive.
❏ Carry out drills involving gas, water, electricity turn-off and activation of emergency generators.

Student/Parent Reunion
❏ Work with the school principal to establish a policy for all employees which addresses both school and personal needs.
❏ Develop procedures that specify how staff will handle a release, in view of available damage information — for the site as well as for the larger community.

Fire Suppression
❏ Make sure that extinguishers are in working order and that other equipment is complete and in easily accessible places.
❏ See that all staff receive training in equipment use and in how to notify fire departments and police departments.

Search and Rescue
❏ Make sure necessary supplies (crowbars, hard hats and gloves) are on-site and readily accessible.
❏ Make sure staff members stay current with their training.

First Aid
❏ Make sure that first aid supplies are up-to-date and always complete.
❏ Keep emergency cards (lists of medical resources in the area) and health cards (for each employee and student) up-to-date.
❏ Develop a method of direct communication between all areas of the school and the command post.

Evacuation (This may be a Function of the Emergency Management Team)
❏ Keep plans for designated emergency assembly area current.
❏ Make sure necessary supplies are accessible.
❏ List those students who will need assistance in the event of an evacuation, and develop a plan to assist and assign staff or other students to help specific individuals.
❏ Perform practice drills.

Damage Assessment
❏ Identify team members who are familiar with the school.

❑ Develop a plan to inspect the school in an orderly manner.
❑ Gather floor plans and maps, as needed.

Note: Because of high staff and student turnover, conduct orientation, equipment supply checks and drills quarterly, if not more frequently.

■ *Address the language needs of non– or limited-English speaking students and staff in preparedness planning, training and drills. The American Red Cross, for instance, can deliver materials in several languages other than English, including Spanish.*

BEFORE HELP ARRIVES

The Emergency Checklist

There is, of course, a possibility that at the time of the event, the person in authority at the site may be confused as to what steps to undertake first. The following **Emergency Checklist** suggests **prioritized** procedures to follow.

1. Go to the command post.
2. Turn on a portable radio.
3. Round up all flashlights/light sticks and extra batteries.
4. Activate the Incident Management System (IMS).
5. Provide a briefing based on the information at hand.
6. Test telephones and walkie-talkies.
7. Determine incident priorities, and begin to act based on these priorities.
8. Schedule meetings at regular intervals to share updated information.

Emergency Procedures

DIRECTION AND CONTROL

Function: Command Post (School Principal or a Designated Representative)

❑ Activate the command post.
❑ Keep a record of events, decisions and actions.
❑ Account for all employees, students and visitors.
❑ Implement and coordinate emergency operations.
❑ Request situation and damage reports from all emergency response personnel.
❑ Develop and display situation status.
❑ Determine whether evacuation is necessary, and communicate that decision to all employees and students.
❑ If there is the slightest suspicion that the school has suffered structural damage, contact the architect or structural engineer, with whom the school has a pre-existing agreement, to conduct a post-disaster inspection.
❑ Continue to update situation information as additional emergency response reports come into the command post.
❑ Control internal and external communications.

Function: Support & Security

❑ Check utilities and do whatever is necessary to minimize any danger. Determine what utilities still work and what does not. Report these findings to the command post.

❏ Make a note of structural and non-structural damage when checking utilities. Report any identified damage to the command post.

❏ Assist in evacuation, if required.

❏ Set up an emergency sanitation system or procedures, although be sure to avoid the use of water or toilets until after a check of lines for breakage.

❏ Monitor the use of emergency water supplies (including water from hot-water heaters).

❏ Inventory supplies of available food. Begin to plan the distribution of food, if the situation warrants. Station someone at the main gate/front door to deal with the media and the community. Route fire, police, rescue, medical personnel and volunteers to the necessary area, and keep the command post informed of activities.

Function: Fire Suppression

❏ Initiate a response.

❏ Check for and confirm the existence of fire(s), and report locations to the command post.

❏ Control fire, if possible.

❏ Notify the fire department, as necessary.

❏ Rescue anyone at risk.

❏ Secure areas.

Function: Search and Rescue

❏ Initiate response.

❏ According to pre-established pattern, check (visually, vocally and physically) every room, and report problems to the command post.
❏ Assist in the administration of first aid, as appropriate.
❏ Conduct a sweep through the school, and look for obvious structural problems or significant structural damage. Report any damage to the management center.
❏ Assist students who use wheelchairs and walkers, if evacuation is necessary.
❏ Lead students with dementia out of the building, if evacuation is necessary.

Function: First Aid
❏ Initiate response.
❏ Make a situation report immediately to the command post.
❏ Administer first aid. Record all cases and treatments.
❏ Determine the need for further medical assistance, and coordinate requests for assistance through the command post.
❏ Inventory supplies of available food, and begin planning food distribution, if the situation warrants.

Function: Damage Assessment
❏ Inspect the school.
❏ Take photographs, or videotape, any damage.

- ❏ Keep detailed records. Document purchases and repair work. Keep all receipts.
- ❏ Contact the school's insurance company.
- ❏ Contact a debris removal company.
- ❏ Conduct salvage operations. Separate damaged from undamaged property.
- ❏ Take an inventory of damaged property. Keep damaged property and goods on hand until the insurance adjuster assesses the damage.
  Make temporary repairs to protect undamaged property.
- ❏ Assess remaining hazards and maintain property security.

Function: Student Staff Supervision
- ❏ Prepare to relieve a buddy teacher of their students.
- ❏ Take charge of multiple classrooms.
  Inform students of the crisis, as the information arrives.
- ❏ Help students talk about the emergency.

Function: Student/Parent Reunion
- ❏ Turn students over to a buddy teacher.
- ❏ Report to the command post, and join up with other team members.
- ❏ Obtain necessary supplies, and establish parent relocation access points.

RESPONDING IN THE AFTERMATH OF CRISIS

Members of the crisis team should understand natural stress reactions. They also should be familiar with how different individuals might respond to death and loss, including **developmental considerations**, **religious beliefs** and **cultural values**.

Effective schools will ensure a coordinated community response. **Professionals** both within the **school district** and within the **greater community** should assist individuals who are at risk for severe stress reactions.

Schools that have experienced tragedy have included the following provisions in their response plans:

♦ **Help parents understand their children's reactions to violence.** In the aftermath of tragedy, children may experience unrealistic fears of the future, have difficulty sleeping, become physically ill, and/or become easily distracted (these are common symptoms).
♦ **Help teachers and other staff deal with their reactions to the crisis.** Debriefing and grief counseling are just as important for adults as they are for students.
♦ **Help students and faculty adjust after the crisis.** Provide both short-term and long-term mental health counseling following a crisis.

♦ **Help victims and family members of victims re-enter the school environment.** Often, school friends need guidance in how to act. The school community should work with students and parents to design a plan to make it easier for victims and their classmates to adjust.

♦ **Help students and teachers address the return of a previously removed student to the school community.** Whether the student is returning from a juvenile detention facility or a mental health facility, schools need to coordinate with staff from that facility to explore means to make the transition as uneventful as possible.

CRISIS PROCEDURE CHECKLIST

A crisis plan must address many complex contingencies. There should be step-by-step procedures when a crisis occurs:

❏ Assess life/safety issues immediately.
❏ Provide immediate emergency medical care.
❏ Call 911 and notify police/rescue **first**. Call the superintendent **second**.
❏ Convene the crisis team to assess the situation and implement crisis response plans.
❏ Evaluate available and needed resources.
❏ Alert school staff to the situation.
❏ Activate the crisis communication procedure and system of verification.
❏ Secure all areas.
❏ Implement evacuation and other procedures to protect students and staff from harm.
❏ Avoid dismissing students to unknown care.
❏ Adjust the bell schedule to ensure safety during the crisis.
❏ Alert persons in charge of media outlets to prevent confusion and misinformation.
❏ Notify parents.
❏ Contact appropriate community agencies and the public information office.
❏ Implement post-crisis procedures.

Chapter VI: Transportation Systems

Introduction

Transportation Infrastructure

Security Issues and Transport Systems

Response to a Crisis, Attack or Disaster

Initial Assessment

Target Folder/Response Information Folder

Terrorism and Attacks Against Transit Systems

Recording Threats

Searches

Incident Objectives for Terrorist Attack

Scene Security Concerns

CBRN Incident Indicators

Transportation Systems

INTRODUCTION

This chapter provides an overview of security and terrorism issues that affect transportation infrastructure. While these issues also impact freight systems, this chapter focuses on those systems that move people: *passenger rail systems* (subways or metros, light rail, commuter and inter-urban rail), *transportation terminals* (bus, rail, ferry and airport terminals) and *ferries*. Transportation security issues include crime, crowd management, disruption of the transportation infrastructure and attacks against the system that involve terrorism or sabotage. The chapter will also address attacks involving Weapons of Mass Destruction (WMD), such as Chemical, Biological, Radiological or Nuclear (CBRN) terrorism in the transport environment, and include checklists for transit incident management.

TRANSPORTATION INFRASTRUCTURE

Transportation infrastructure includes both **publicly– and privately-owned passenger and freight transport assets**. These include various modes of transportation including railways, buses, airlines, vessels (ships and ferries) and their terminals: airports, seaports, rail terminals and stations. Transport of both people and goods is a critical element of modern life, and forms an integral component of communities and their economies.

Connectivity between communities, individual mobility, and the ability to transport goods and people are essential capabilities and services. **A disruption of these services can place communities and national security at risk**. Even short-term, localized disruption of transportation can result in **significant economic losses and an interruption of vital services**.

Individual transport systems or modes are not only vulnerable to short-term disruptions due to **natural disasters** (such as extreme weather events, or earthquakes), **accidents** or **intentional attacks** (such as terrorism or sabotage). Disruptions in the operations of other infrastructures can also impact transportation networks. Transport is increasingly dependent upon the **energy, communications**, and **information infrastructures**. Understanding these interdependencies and vulnerabilities is a critical element in a comprehensive transport security strategy.

SECURITY ISSUES AND TRANSPORT SYSTEMS

The importance of public safety and security are common to all transport modes. Persons using transport systems expect a comfortable, safe and secure travel environment where they can travel without fear of violence.

Accordingly, transport operations and security personnel must recognize the importance of **passenger security issues** to the on-going success of their transport system. The maintenance of **order**, **crime prevention**, **minimal disruptions and delays**, **minimal civil liability**, and **crowd/passenger management** are elements that are common to all transport modes.

Key issues in the rail setting include: crimes such as vandalism; graffiti; passenger robberies; fare evasion; assaults against passengers and transit personnel; and critical incidents including:

♦ Fires/smoke conditions.
♦ Grade crossing accidents.
♦ Persons struck by trains.
♦ Natural disasters/extreme weather events.
♦ Train collisions/derailments.
♦ Bomb threats, sabotage and terrorism.

Bus systems share many concerns with their rail counterparts, including crowd management during special events, alcohol-related incidents, and assault/robbery of transit personnel and/or passengers. **Ferry systems** are experiencing a resurgence of importance in congested urban areas along harbors, rivers and lakes. **Passenger ferries** and **water taxis** have vessel-specific concerns that pose special security challenges to law enforcement personnel. One example is **peak-period crowd surges**, when large numbers of passengers attempt to board vessels. As passengers push and shove, they can cause a safety hazard at the wharf-vessel interface. Similar issues can occur at **disembarkation** points. Incidentally, both periods are prime periods for **pickpockets** to act, for **thieves** to snatch purses and chains, for disturbances to develop and for **terrorists** to stage attacks.

Key ferry security issues include personal safety at the wharf or terminal (slip), peak period crowd control, on-vessel safety, disembarkation, and personal safety en-route to a following travel mode (**car**, **bus** or **train**). While robberies or assault aboard vessels can occur, authorities can usually thwart suspects from escaping, since transit personnel can keep vessels on the water pending the arrival of police.

RESPONSE TO A CRISIS, ATTACK OR DISASTER

Management and response to critical incidents (crisis, attack or disaster) are demanding tasks. Disasters include situations that require a significant application of external resources to effect a resolution. **Crisis** refers to the incident's impact on the organization, as well as the organization's ability to cope with or respond to an extraordinary incident or event. **Responders** and **command personnel** in these situations may face a variety of demands in an extremely short time frame, in stressful and sometimes dangerous or austere conditions. An understanding of what is happening (e.g. the situation that personnel face) is commonly known as **situational awareness**. Certain common attributes of both **disasters** and **crises** that can complicate decisionmaking include:

♦ Uncertainty.
♦ A lack of timely information.
♦ Conflicting information.
♦ Fog, friction and noise.

In this context **fog** refers to the obscuring of situational awareness, and **friction** to unexpected factors that complicate response. In these situations, the simple becomes hard, and the hard becomes harder. **Noise** refers to an excess of non-relevant information. Furthermore, management of complex events entails risk (both organizational and personal) and can have severe consequences. **Time constraints and incomplete or conflicting information can often complicate complex management**. In terrorist attacks, the presence of an opposing will further complicates the situation.

In order to overcome these inherent disadvantages, management should employ **preparation**, **recognition** (situation assessment and size-up, key indicators), and **management** tools (response objectives, courses of action, incident action plans).

A size-up or situation estimate essentially establishes: **1)** what is happening?, **2)** what do I have to do to deal with it? (using immediately available resources), and **3)** what resources are reasonably available?

Once planners have answers to these questions, they should request necessary resources and prioritize needs. A short checklist for conducting an initial assessment and initiating response follows:

INITIAL ASSESSMENT (SIZE-UP/SITUATION ESTIMATE)

♦ Type of emergency.
♦ Location (site and distinguishing characteristics).
♦ Size of the involved area.
♦ Number and type of casualties.
♦ Special hazards.
♦ Assistance required (number of police officers, transit vehicles, utilities, etc.)

Initial Response Considerations

1. Assess the situation (size-up/situation estimate).
2. Communicate the assessment to dispatchers/ incoming units.
♦ Include identification of responder hazards.
3. Provide direction to incoming units.
♦ Safe approach, safe staging/ mobilization areas, PPE needs.
4. Establish containment.
♦ Preserve the scene, inner/outer perimeter, and zones of operation.
5. Determine the need for immediate protective measures.
♦ Turn off ventilation systems.
♦ Evacuation, in-place protection, or a mixed strategy.
6. Select and communicate the command post location.
7. Establish the command (ICS/unified command structure).

8. Develop incident objectives and an action plan.
9. Request appropriate staffing and equipment.
10. Remember traffic control, crowd control and force protection.
11. Make appropriate notifications.
♦   If management suspects terrorism or CBRN, notify the FBI.

A formal process for the development of **Courses Of Action (COAs)** helps to guide response activities in complex situations. A COA should rest upon situational awareness, and reflect a clearly defined mission bounded by a commander's intent. It must explicitly state **constraints** (what must be done) **restraints** (what must not be done) and state **assumptions** on which the actions are based (in case they change). COAs support the **Incident Action Plan (IAP)** or operation plan. An IAP should define objectives, options (strategy) and strategies. An IAP is subject to on-going size-up and refinement, and should contain checklists/matrices to encourage understanding by end users.

■ *An IAP is an essential element in the planning and documentation of the actions that will be necessary to effectively resolve an incident.*

The development of response checklists and site assessments can facilitate pre-planning endeavors to manage complex situations such as disaster or terrorist incidents. This process is known as **Intelligence Preparation for Operations (IPO)**. Key IPO products include a pre-designated command post, staging, decon, casualty collection and rendezvous points, as well as the development of **target folders** (or response information folders) to facilitate response. The contents of a typical target folder follow:

TARGET FOLDER/RESPONSE INFORMATION FOLDER

- ♦ Location.
- ♦ Type of facility.
- ♦ Daytime/night-time population.
- ♦ Points of contact (POCs).
- ♦ Voice/fax/e-mail/websites.
- ♦ Unique hazards.
- ♦ Past threat history.
- ♦ Key dates.
- ♦ Floorplans.
- ♦ Photos (ground level aerial, key exits, hazards, etc.)
- ♦ Heating/ventilation system characteristics.
- ♦ Blast analysis.
- ♦ Interior/exterior plume dispersal (for CBRN or Haz-Mat events).
- ♦ Downwind or downhill potential (collateral impact).
- ♦ Communications capabilities/limitations.
- ♦ Available power, water and lighting.
- ♦ Secondary/systemic impacts.
- ♦ Response resources (fire, EMS, HazMat).
- ♦ Pre-designated support facilities (CP, staging, decon, etc.)

TERRORISM AND ATTACKS AGAINST TRANSIT SYSTEMS

In recent years, transit or transport systems of all types have been subject to terrorist attacks. **Hijackings** and **aircraft bombings** have been the primary transport types of attack, and these events have generally received the most public attention.

Recently, however, worldwide attacks have also targeted **surface transport systems**. Historical transit attacks include the Fulton Street firebombing on the New York City Subway in 1994; the derailment of Amtrak's Sunset Limited, in Hyder, Arizona in 1995; the Long Island Rail Road shooting, in which a lone gunman killed six and injured 20 in 1993; and the Tokyo sarin gas attack, in which Aum Shinrikyo killed 13 and injured up to 5,500 others in 1995.

Attacks can include **sabotage** or use of **bombs**, **chemical** or **biological agents**, **nuclear** or **radiological materials** or armed assault with **firearms** or other weapons by terrorist or quasi-terrorist actors. These events may cause substantial damage or injury to persons or property.

Recent technological advances make it necessary to add assaults on transportation information systems (**virtual attacks, cyber-attacks** or **cyber-terrorism**) to the list of types of attack. Virtual attacks or attacks using an information system yield a **physical result**, while cyber-attacks are attacks against an **information system** itself. Hybrid attacks employ more than one attack type, usually a combination of a conventional attack (**bombing, armed assault**) in combination with a **chemical, biological, radiological** or **cyber-attack**. Targets of these kinds of attacks can include all modes of transport such as **aircraft, rail systems, buses, vessels,** and **terminals** of all types.

Attack types include:
♦ Armed assault.
♦ Explosives.
♦ Arson (incendiary devices).
♦ Chemical agents.
♦ Biological agents and toxins.
♦ Nuclear/radiological materials.
♦ Virtual and cyber-attack.

Terrorism as Intentional Disaster

Transit systems, including transportation vehicles and terminals of all types, are increasingly potential targets of terrorism. Transit systems have several common characteristics that render them particularly vulnerable. A summary of these characteristics follows:

♦ Transit systems carry large numbers of people within enclosed spaces.
♦ Transit systems follow known routes, and usually pass at predictable times.

♦ Transit systems have fixed points of egress and ingress.
♦ There are already unique hazards associated with transit systems (e.g. high voltage traction power) which can complicate first response.
♦ Transit systems are subject to systemic impact: an attack at one point will often impact other remote segments of the system.

Terrorists (or people who employ terrorist-like tactics — quasi-terrorists) can exploit the vulnerability of a transit system to **1)** conduct direct attacks aimed at damaging the system or generating casualties; or **2)** utilize threats or acts to stimulate disruption. Within the US, **threats are far more common than actual terrorist or quasi-terrorist attacks**. Management of threats and actual terrorist events requires the maintenance of on-going security programs that can guide response capabilities. Key steps in threat management follow:

Key Threat Management Steps
♦ Review threat information.
♦ Establish a decision authority (this is the person responsible for determining a course of action in response to a threat).
♦ Conduct a threat/risk evaluation.
♦ Determine the potential impact to people and the system.

♦ Assess system status (location of transit vehicles, passengers).
♦ Develop an Action Plan (AP).

Threat Management/Bomb Threats

Bomb threats are common features at many transit systems. If management receives a threat, they must record and review the threat. Management must make a decision on what to do, based on the credibility of the threat, its technical feasibility, and its potential impact on **resources/system** and **people/passengers**. This could include:

♦ No action/normal service.
♦ Restricted service/localized search or evacuation.
♦ Suspension of a service/system-wide search or evacuation.

Finally, develop an action plan, implement a response organization and make appropriate notifications. If management decides to take action, define the incident area (incident on a train, at a station, or in a terminal area, etc.) Monitor CCTV, if available. Relay passenger (PAX) load and status, in the affected area and system-/facility-wide, to the incident commander.

RECORDING THREATS

Upon the receipt of a bomb threat (or other threatened action, such as an anthrax threat, a sabotage threat, etc.), record it to facilitate threat assessment. Record all threats and keep all threats on file. This will ensure comprehensive threat management. Facility management will be able to contrast current threats with past threats, to aid in analysis. The following summarizes the actions that management can take, and questions worth asking and recording.

Actions to be Taken upon Receipt of a Telephonic Threat
♦ Switch on a voice recorder, if available.
♦ Record the exact wording of the threat.
♦ Ask these questions:
   1. Where is the bomb (device) right now?
   2. When is it going to explode (activate)?
   3. What does it look like?
   4. What kind of bomb (device/agent) is it?
   5. What will cause it to explode (activate)?
   6. Did you place the bomb (device)?
   7. Why?
   8. What is your name?
   9. What is your address?
   10. What is your phone number?
♦ Record the time when the call ended.
♦ Record caller's phone number from 'caller ID', if available.

♦ Inform police and the site security coordinator.
♦ Record the gender, ethnicity, accent, speech characteristics, nature of the threat language used, background sounds, and any other telling information or remarks.
♦ Utilize prepared threat forms, if available.

SEARCHES

If management elects to conduct a search, the **incident commander** must safely manage and resolve the incident, maintain an effective perimeter, make effective use of all resources, establish and operate an effective command post, and remain aware of other targets and secondary impacts on the transport system. The effective exchange of information, as well as interagency communication, are both essential.

All searches are **visual** surveys. No searcher should touch anything. The search should follow a plan, with defined objectives and clearly articulated search parameters. Teams should conduct systematic and coordinated searches, and there should be a manageable span of control. **In the event that teams locate a device or suspect package, they should immediately leave the area, and management should safely isolate and deny entry to the area. All searchers should immediately clear the area**.

Isolate the area to a distance of at least 500 feet, in every direction.

Clearly enumerate safety objectives **prior** to the initiation of any search. Typical objectives include:

♦ Use no radios or cellular phones (or other electronic devices) in a suspect area. Such devices can set off certain detonators.
♦ Employ safe search practices.

- ◆ Advise all incident personnel of safety issues.
- ◆ Safely isolate and deny access to an area if search teams locate a device (bomb) or a suspect package.
- ◆ Clear, isolate and label areas where personnel have already conducted searches, without locating devices.
- ◆ Maintain an awareness of routine hazards (train operations, trip hazards, moving vehicles, etc.)

INCIDENT OBJECTIVES FOR TERRORIST ATTACK

The following table summarizes typical incident objectives for response to a terrorist attack. It includes both general and transit-specific objectives:

General Objectives
♦ Secure perimeters.
♦ Control and identify the threat.
♦ Rescue, decon, triage, treat and transport injured persons.
♦ Move crowds to safe zones.
♦ Protect rescuers.
♦ Avoid secondary contamination.
♦ Secure evidence and the crime scene.
♦ Protect against secondary attack.

Transit-Specific
♦ Provide alternative modes of transport (e.g., bus-bridges).
♦ Assess and mitigate secondary impact on the system.
♦ Restore service quickly.
♦ Restore passenger confidence.
♦ Restore employee confidence.

SCENE SECURITY CONCERNS

A terrorist attack at a transport venue requires a variety of law enforcement and security missions. As well as **investigation** and **crime scene** duties, other essential duties include **traffic control**, establishment of a **cordon** (inner and outer perimeters), and **force protection** (measures to protect response personnel). Key force protection missions include:

♦ Scene security.
♦ Protection of responders, equipment and facilities.
♦ Awareness of secondary attack.

The potential for **secondary attack** is a major security consideration. Recent trends in terrorist attacks indicate a need for greater vigilance in this area. The predominant mode of secondary attack is a **secondary device** in bomb scenarios.

For example, terrorists may place a bomb on the incident ground in order to injure responders. Response to potential terrorist attacks should include provisions for screening **incident support areas** (command posts, staging areas, decon corridors, etc.) for secondary devices. Coordination with **explosive ordnance disposal (EOD)**/bomb squad personnel is also necessary. If responders locate a **secondary device**, personnel should immediately leave the area, proceed to a safe zone, and deny access to the hazard area.

All response personnel should be familiar with procedures for addressing secondary devices or secondary attacks. An **immediate action drill** is one method to ensure an appropriate reaction. The following is an immediate action drill for secondary attacks, based on the mnemonic, **RESCUE**:

Secondary Attack: Immediate Action Drill

R – Repel the threat.
E – Extract personnel and injured people.
S – Safe area (remove personnel to a safe area).
C – Control team (control the actions of personnel).
U – Urgency of care (determine the status of injured personnel, treatment needs and priorities).
E – Evacuate (evacuate the injured to definitive care).

Chemical, Biological and Radiological Attacks and Indicators

Terrorists may employ chemical, biological or radiological weapons against transportation systems, seeking to cause mass casualties, mass disruption or to heighten fear and terror. Transit systems, particularly **terminals**, **vehicles** and **vessels**, concentrate a large number of people in a compact space. CBRN events constitute intentional hazmat scenarios that require the establishment of a crime scene and a multi-jurisdictional/multi-disciplinary response. **Casualties** (persons injured or dead) may result and **mass decontamination** (decon) may be necessary, requiring the management of existing transit-specific hazards.

■ *The recognition of WMD events is more complex than the recognition of explosive incidents. Each type of agent, whether chemical, biological or radiological, possesses unique characteristics.*

Chemical attacks are more predictable in their results and generally yield an immediate impact. **The Tokyo sarin gas attack is a classic example of a chemical attack against a transit system.** During the Cold War, Soviet, American and UK military and security services assessed the potential of a biological attack (**anthrax**, for example) against the Moscow, New York and London subway systems, respectively. Each nation considered subways viable targets for attack involving aerosol dissemination.

Biological attacks are even more difficult to discern, since their respective incubation periods may mask the presence of certain biological agents. Many analysts believe that experts will initially mistake bio-attacks for suspicious outbreaks of disease. A biological attack takes hours (in the case of **toxins**) to days or weeks to yield effects. **This makes immediate recognition problematic unless experts receive a threat or discover direct evidence, such as the discovery of dispersal devices**. Persons infected by a biological attack will likely become conscious of illness well after they leave the focal point of the incident area.

Nuclear attacks are far less likely, and would reveal themselves through a large blast. On the other hand, radiological attacks that involve **radiological dispersal devices (RDDs)** would be largely tools of disruption. **Radiation is invisible, odorless and tasteless**, and this would likely mask the discovery of a radiological attack, and cause a delay in the onset of radiological symptoms. Unless a facility or responders employ radiological survey instruments (detectors), the recognition of a radiological attack would be dependent upon the discovery of direct evidence (RDD or radio-luminescent material) or the receipt of a threat communication from a terrorist.

■ *Following an attack, it is important to ensure that the transport system is free from residual contamination, and that the public accepts this determination. Perceptions of safety will affect public confidence and future transit use.*

The following are checklists to ensure that public safety, security and law enforcement personnel who work at or respond to special events are able to recognize a chemical, biological, or radiological incident. **These elements are essential in pre-event planning and preparation**. The following checklists summarize the indicators of chemical, biological and radiological incidents for response personnel and commanders:

CBRN INCIDENT INDICATORS

Chemical Incident Indicators
Minutes to hours...

Unusual dead or dying animals
— Lack of insects
Unexplained casualties
— Multiple victims
— Serious illnesses
— Nausea, disorientation, difficulty breathing and convulsions
— Definite casualty patterns
Unusual liquid, spray or vapor
— Droplets, oily film
— Unexplained odor
— Low-lying clouds/fog unrelated to weather
Suspicious devices/packages
— Unusual metal debris
— Abandoned spray devices
— Unexplained munitions

Biological Incident Indicators
Hours to days...

Unusual dead or dying animals
— Sick or dying animals, people or fish
Unusual casualties
— Unusual illness for the region/area
— Definite pattern inconsistent with natural disease

Unusual liquid, spray or vapor
— Spraying and suspicious devices or packages
Unusual swarms of insects
— Vectors
Suspicious outbreak of disease

Radiological Incident Indicators
Delayed onset of symptoms...

Unusual numbers of sick or dying people or animals
— Symptoms of radiation exposure
Unusual metal debris
— Unexplained devices/munitions-like material
Radiation symbols
— Placards on container
Heat emitting material
— Emitting heat without signs of an external heating source
Glowing material/particles
— Radio-luminescence

Chapter VII: Utilities

Introduction

Utility Companies as Targets

Step One: The Planning Process

Step Two: Planning Checklists

Step Three: Response Folders

Step Four: Emergency Tasks

Terrorist Weapon Systems

Utilities

INTRODUCTION

The utility industry comprises an array of critical systems and networks. **Electrical**, **water** and **communications** utilities and **pipelines** (gas and oil) are all vital nodes in society.

Terrorists do not need to employ weapons of mass destruction to execute a massive strike against **infrastructure**. Terrorists have the capacity to apply a crude but effective explosive device at a critical juncture, and thereby cripple a section of a city or even a sector of an entire country, **whether in highly industrialized or developing countries**. These types of scenarios can also enable terrorists or criminals to create a diversion, and sow confusion as a prelude to an attack against their real objective.

UTILITY COMPANIES AS TARGETS

Utility companies have long been worried about terrorist attacks. Gas and oil pipelines, as well as electrical power, water supply and communication systems are all vulnerable networks. This has made them increasingly attractive terrorist targets.

Electric Utilities

The vulnerability of electric power grids has recently caused them to rise to the top of the list of terrorist targets. Power grid design includes **loops**, which allow a grid to sustain numerous power outages and failures, and still provide electricity service to customers. Yet terrorists comprehend that this loop characteristic of power grids works to their advantage.

Terrorist attacks on select points of a power grid can lead to **widespread blackouts**. In one outage that occurred in the Western US in 1996, for example, an overloaded system at a single choke point caused a blackout over eight states for a period of hours. This form of attack constitutes **sabotage**, but in this situation terrorists would be in a position to then carry out other planned attacks in other affected regions.

Gas and Oil Utilities

Also vulnerable are oil and gas pipeline systems, both over land and on the water. Terrorists can perpetrate acts clandestinely against **platforms** that lie off a coastline, through means such as suicide bombers who pilot pleasure craft, scuba divers who can affix mines or explosives to a platform base, or kamikaze-style aircraft.

On land, **gas and oil pipelines** are highly susceptible to attack. Safety precautions ensure that pipeline operators clearly label/identify their pipe and product every time it traverses a road or waterway. What is logical from the standpoint of safety, however, has the potential to enable terrorists to **clearly identify and target** pipes that carry oil, natural gas or volatile chemicals.

In addition, the vulnerability of **tanker trucks** and ships is just as high. A liquid petroleum gas (LPG) tanker car with a 35,000-gallon capacity has an explosive potential of about 0.75 kilotons, while an ocean-going **supertanker**, carrying liquid natural gas (LNG) can generate a large-scale explosive plume and massive firestorms, whose temperature could exceed 3,000° F (1,649° C). A full supertanker can yield a TNT equivalent of 0.7 megatons. The possible ramifications of explosions on this scale lay bare the vulnerability of gas and oil utilities, and their associated transportation systems, to terrorist attack.

A chapter on **Utilities Security** could include volumes of information, if it treated all types of attacks, using all types of weapons. This chapter will rather provide planners with the analytical tools to effectively:

♦ Begin to employ a proactive planning methodology.
♦ Build **response folders**, which are critical documents for emergency planning and response.
♦ Establish an incident response structure.

STEP ONE: THE PLANNING PROCESS

The goal of a "living" planning guide is to provide the planner with the information and necessary tools to plan, prepare, respond to, and then integrate planning with all public safety organizations that will play a role in the event of an attack on a facility.

Utility providers in a given area probably will most likely already have pre-existing relationships with members of the **public safety community**. The assembly of a proper plan, and effective coordination with relevant safety organizations will ensure **joint agency integration**, and the efficient utilization of **resources** during a developing terrorist or criminal incident.

STEP TWO: PLANNING CHECKLISTS

1. Developing an Emergency Operations Plan (EOP).

It will become clear in the early planning stages that most **Emergency Operations Plans (EOPs)** have a similar layout. Typical plans assign responsibility to an organization or to individuals, in order to carry out specific actions at projected times and places during emergencies. **An EOP sets organizational relationships and authority guidelines**. It also details how responders will protect people and property in a given situation. An EOP will describe how facility personnel will integrate and utilize equipment, personnel, resources and supplies during an event. It will establish authorities for particular functions during an emergency. Finally a good EOP will address **mitigation strategies** following an event.

2. Conduct a Vulnerability Assessment of the Facility

Develop a list of vulnerabilities from local resources or from the attached list (Table I). Seek the advice and assistance of the planning team to ensure the development of the most comprehensive list possible, to better prepare for an attack.

		Table I: Sample Vulnerability Assessment
❏	1.	Minimal security.
❏	2.	Easy access to the public.
❏	3.	Mixed business interaction with minimal contact. An example of this might be a clinic or office that operates in a location away from the secure hospital and has no reception point. This provides a terrorist the ability to conduct surveillance and gain intelligence without being challenged.
❏	4.	Central receiving. Personnel assigned to central receiving may be less likely to question suspicious packages because they are unfamiliar with the size and types of packages the organization receives.
❏	5.	Multilevel buildings. Facilities with many levels present a challenge to all planning documents. First, they are difficult to evacuate in any situation. Second, they provide an opportunity for a terrorist to hide devices or themselves. Third, elevator shafts and stairwells can carry contaminants through a building while individuals attempt to evacuate.
❏	6.	Groups in an area.
❏	7.	Parking garage.
❏	8.	Restricted exits/fire escapes.
❏	9.	Inadequate air handling systems.
❏	10.	No control of HVAC or other utilities in the building.
❏	11.	On site storage of HazMat and Bio-waste material.
❏	12.	Close proximity to other industrial businesses.
❏	13.	Routes of passage for hazardous material in close proximity. Does a major trucking route pass directly in front of the location?

Table I: (continued)

❏	14. Easy escape route via major roadway.
❏	15. Concealment locations in close proximity. This could allow an attacker to hide his/her location in an attack and to execute a secondary attack.
❏	16. No visible security presence.
❏	17. Routine response to unscheduled events. Do personnel take events for granted? Do they respond to "bomb hoaxes" with reluctance?
❏	18. Do personnel challenge unscheduled or unannounced visitors?
❏	19. Has there already been an attack or incident of some kind?
❏	20. Does the organization sponsor or support an event that may be controversial?
❏	21. Is operational capacity beyond the abilities of security personnel?
❏	22. Is the organization impenetrable?
❏	23. Are personnel members learned in surveillance methods?
❏	24. Is there a reporting method for the organization regarding unusual events?
❏	25. Does a management team routinely inquire about strange occurrences at the organization's facilities or locations?
❏	26. Is there intelligence coordination with other utility providers in the area?
❏	27. Have there been acts of sabotage or theft by unknown individuals or disgruntled employees in the past?

After the completion of a vulnerability list, write the questions in one column (as described in **Chapter IV: Hospitals**). The planning team should conduct an assessment using the vulnerability list. Place a check in the column applicable to the target. Total check marks in each of the columns. The higher the column total, the more susceptible is the target to a terrorist attack (see Table II for a sample utilities checklist).

Table II: Sample Planning Checklist		
Location Name (eg Refinery)	Location Name (eg Power Station)	
✓	❏	Minimal security.
❏	✓	Easy access to the location.
❏	❏	Central receiving.
✓	❏	Multilevel buildings.
❏	✓	Groups in an area.
❏	✓	Parking garage.
❏	✓	Restricted exits/fire escapes.
❏	✓	Inadequate air handling systems.
❏	✓	No control of HVAC or other utilities in the building.
❏	✓	On site storage of HazMat and Bio-waste material.

		Table II: (continued)
❏	❏	Close proximity to other industrial businesses.
✓	❏	Routes of passage for hazardous material in close proximity.
✓	❏	Easy escape route via major roadway.
✓	❏	Concealment locations in close proximity.
❏	✓	No visible security presence.
❏	✓	Routine response to unscheduled events.
❏	✓	Do personnel challenge unscheduled or unannounced visitors?
✓	❏	Has there been an attack or incident of some kind?
✓	❏	Does the organization sponsor or support event(s) that may be controversial?
✓	❏	Is operational capacity beyond the abilities of security personnel?
❏	❏	Is the organization impenetrable?
❏	✓	Are personnel members learned in surveillance methods?
❏	✓	Is there a reporting method for the organization regarding unusual events?
✓	❏	Does a management team routinely inquire about strange occurrences at the organization's facilities or locations?
✓	❏	Is there intelligence coordination with other utility providers in the area?

		Table II: (continued)
❏	✓	Have there been acts of sabotage or theft by unknown individuals or disgruntled employees in the past?
10	13	**Total**

The planning team can also employ pre-established security questionnaires, which can also prove helpful (see **Chapter II: Facility Security Planning**, for examples).

STEP THREE: RESPONSE FOLDERS

As in earlier chapters, planners and management must develop a process to begin a systematic step-by-step facility planning process. This exercise will provide guidelines to help planners build and plan a **response folder** for a given location. This document will provide planners and response agencies with the necessary information to respond to an incident at a facility.

For utilities, planners should prioritize each individual building/facility according to vulnerability and threat. As such, planners should compile **response folders** in the same manner.

RESPONSE FOLDER CONTENTS:

SECTION ONE
1. General site information:
   a. Role of the facility in the community.
   b. Economic impact on the local area.
   c. Employment.

2. Location:
   a   Detailed strip maps to the location from the nearest airports, interstate highways and other principal routes.
   b. Surrounding area or local area maps (counties, cities and towns within a 30-mile radius).
   c. Maps to additional facilities or support facilities.

Chapter VII: Utilities

3. Location boundaries:
   a. Area.
   b. General data.

4. Facility description:
   a. Photos.
   b. GPS coordinates.
   c. Map references.
   d. Azimuth and distances to local features.
   e. Grounds description.
   f. Helicopter landing zones with approach obstacles (trees, lights, radio towers and wires).
   g. Facility construction.
      i. Roof entrances.
      ii. Roof type and strength.
      iii. Fire escapes and ladders.
      iv. Stairways.
      v. Elevators.
      vi. Loading docks.

5. Security:
   a. Protective measures.
      i. General defensive strategies.
      ii. Protection strategy.
      iii. Security force in facility.
      iv. Security communications.
   b. Restricted areas.
      i. Reason for restriction.
      ii. Security measures in restricted areas.

    c. Security systems.
       i. Intrusion detection.
      ii. Surveillance measures.
      iii. Fire monitoring system.

6. Electrical power sources:
    a. Security system power supply.
    b. Backup power.
    c. Operations power.
    d. Diagrams (panel locations).
    e. Matrix.
       i. Area/equipment affected.
      ii. Breaker.
      iii. Panel.
      iv. Panel locations.
    f. Impact of loss of electrical power.

7. Ventilation systems with photos:
    a. Air-conditioning unit.
    b. Size.
    c. Alarms.
    d. Monitoring.

8. Telephone system:
    a. Panel locations.
    b. Phone numbers to outside lines.
    c. Servicing company and point of contact.
    d. Cell phone availability and location.

9. Medical support:
   a. Location.
   b. Capabilities.

10. Fire protection:
    a. On-site firefighting equipment.
    b. Sprinkler system.
    c. Extinguisher locations.
    d. SCBA capabilities and locations.
    e. Hydrant locations.
    f. Halon capabilities.
    g. Mutual aid.
    h. First-in unit response times.
    i. Established command cost locations.

11. Hazards:
    a. Toxic chemicals.
    b. Biological waste.
    c. Infectious waste.

12. Water:
    a. Potable water location.
    b. Water shut-off locations.
    c. Emergency storage capabilities and storage locations.
    d. Maintenance company.

13. Sewer:
    a. Manhole locations.
    b. Routing diagrams and dimensions.

14. Vehicles:
    a. Types and number usually found.
    b. Locations of parking lots.
    c. Diagrams and capacities of enclosed or underground parking facilities.

15. Staffing:
    a. Key personnel data.

SECTION TWO

1. Building layout — floor diagrams:
    a. Doors.
    b. Interior windows.
    c. Rooms.
    d. CCTV locations.
    e. Other pertinent security information.

2. Operational considerations:
    a. Likely hostage-holding areas.
    b. Potential adversary holding areas.
        i. Fields of fire.
        ii. Concealment of additional weapons or devices.

3. Critical paths:
    a. Critical path diagrams (with photos).
    b. Critical path doors.
        i. List of doors along path with descriptions.
        ii. Door information matrix.
            1. Interior/exterior.

2. Glass size (wire).
3. Width.
4. Height.
5. Thickness.
6. Lock type.
7. Door opening direction.
8. Material.
9. Keys and key control.

In addition to the establishment of a **response folder**, planners should also consider where to place an initial command post. For off-site facilities, establish **primary** and **secondary** locations for the **incident command post**. This methodology will give planners the ability to decrease a utility's profile and minimize the risk of a terrorist attack.

IMPORTANT NOTE
Obviously, a facility's **response folder** is a valuable document for a terrorist or criminal planning an attack. Therefore, planners **must** keep this document under strict control, and make it available **only** to those individuals who have a required security need to review or examine the material within the folder.

STEP FOUR: EMERGENCY TASKS

After they establish an **EOP** and a **response folder**, planners should establish priorities in the event of a terrorist incident. The following list is a starting point to help planners achieve some or all of the required objectives.

- ◆ Notify emergency response agencies.
- ◆ Notify key hospital planning staff.
- ◆ Activate internal emergency response personnel and resources.
- ◆ Secure and isolate the incident area.
- ◆ Account for the safety of the internal and external population (external refers to people or personnel immediately outside of the facility).
- ◆ Establish the incident command post.
  Assess the situation.
- ◆ Begin evacuation or sheltering in place, if required.
  Provide updates to people located throughout the facility (rumor control).
- ◆ Recommend and coordinate personnel protection measures.
- ◆ Provide overall coordination of the incident.
- ◆ Establish a system to handle public inquires.
- ◆ Establish operational periods for staff, and relieve no essential staff, if able.

Zone Control

Regardless of whether the type of terrorist attack is relatively normal (**armed attack, bombing**), or an unusual occurrence (**WMD attack**), planners and responders must establish initial areas for continued safe operation.

Planners must consider both short-term response operations and extended operations that could last hours or days. While planning for zones of operation always establish:

1. Safe ingress to CP and staging locations.
2. Safe egress of emergency vehicles.
3. Control of ingress and egress around the facility staging areas.
4. Evacuation areas and routes.
5. Inner and outer perimeters.
6. Media holding areas.
7. Hot, Warm and Cold Zones for contamination.
8. Decontamination locations.

IMMEDIATE ACTIONS

When first arriving on-scene following a potential incident, the **correct identification** of the nature of the incident is the first step in first response. There may be an initial period of chaos. This will generally begin following an incident, when most people are in some form of shock. It is during this time that responders and planners must quickly take control.

Gain Control Of The Situation (GCOTS)

1. Size-up the situation.
2. Evaluate the situation.
3. Set incident priorities.
4. Determine the potential for further harm, injury or destruction.
5. Establish one incident command post.

During incident size-up, goals should be to provide answers to three questions:

♦ What is the present situation?
♦ What is the predicted behavior?
♦ How will this affect my incident priorities?

■ *If there is no life hazard, rescue situation or fire then there is no reason to risk exposure to emergency response personnel.*

1. Attempt to approach the incident from uphill, and from an upwind direction whenever possible.
2. Determine the direction from which the wind is blowing (from a position near the incident).
3. Determine a route that will allow for an upwind approach to the incident location.
4. Position apparatus upwind and upgrade the incident site, based on shifts in the wind, whenever possible.

Response considerations
♦ Be alert for outward warning signs.
♦ Be alert for detection clues.
♦ Be alert for booby traps and explosive devices.
♦ Resist rushing in, approach from upwind of the target, and stay clear of vapors, smoke and other hazards.
♦ Implement the ICS:
   a. Activate the emergency operations plan (EOP) and call for additional resources.
   b. Notify state and federal agencies.
   c. Expect unified command operations.

It is important during the initial stages of an incident that managers, planners and responders:

1. Establish clear and achievable incident objectives.
2. Continue to conduct an incident size-up.
3. Implement protective actions:
   a. Secure the area.
   b. Establish decon operations, if necessary.
   c. Evacuate the immediate area.
   d. Protect persons downwind.
   e. Establish site control.
   f. Monitor the progress of incident priorities.

TERRORIST WEAPON SYSTEMS

Terrorist weapons training at terrorist training camps comprise comprehensive, all-encompassing programs. Training ranges from small arms use to the acquisition and production of **Weapons of Mass Destruction (WMD)**. Weapons training also covers weapon **maintenance** and **modification skills**, for a variety of weapons that a group might use through the course of its terrorist operations.

Training is also focused on the development and use of various military weapons systems such as **grenades**, **mines**, **mortars** and various **missile systems**, including the Stinger and Light Anti-tank Weapon (LAW).

The following is a survey of weapons that terrorists frequently use during the course of their military operations. It is not an all-inclusive list, but it does treat major categories and weapons that terrorists have tended to use over time.

Technological advances have granted terrorists access to some of the most lethal portable weapons in the world. These terrorist weapons can range from small, semi-automatic handguns to some of the most powerful explosives and lethal toxins known to man.

Procurement and **construction** of such weapons ranges widely. "How-to manuals" on the Internet outline bomb construction in great detail, while **theft** and **illegal arms networks** around the world offer other avenues for the distribution of effective weapons to terrorists.

Explosives

Terrorists generally employ two types of explosives. These include **fragmentation** and **incendiary devices**. Although such devices can take many different forms (e.g. **pipe**, **letter**, **package** or **vehicle bombs**) they produce similar effects.

All bombs share four standard components. The **first** is the explosive charge. The **second** is a device to initiate the explosion (e.g. the blasting cap). The **third** is a power source to start the process. The **fourth** is a control process.

Military explosives are the most popular explosive charges that terrorists employ. First, they offer greater shattering effects than do commercial explosives. They also have higher detonation rates, and they are relatively **insensitive to heat, shock, impact and friction**. Thus they are safer to handle. Military explosives will work underwater and they are of a convenient size, shape and weight. The most commonly available military explosive is **TriNitroToluene (TNT)**.

Another explosive charge employed widely by terrorists is **Composition 4**, better known as **C4**. C4 is "plastic explosive". Its primary characteristics include a yellowish, putty-like texture, resembling children's Play-Doh.™ One of the most attractive qualities of C4 is the ability to mold it into a variety of shapes. It has an extremely high rate of detonation and a high shatter effect.

Another commonly used military explosive is **PentaeryThriol-TetraNitrate (PETN)**. PETN, also known by the names **detacord**, **primex** and **primeacord**, resembles common clothesline. It has a high explosive charge, and dual military and commercial applications. It is available in linear form and, similar to C4, it is resistant to heat, shock and friction.

Semtex, however, is the favorite high explosive charge for the majority of terrorists. It has characteristics similar to C4 and PETN, as well as high detonation rates and increased shattering effects. Terrorists worldwide also have access to Semtex. For example, *The Washington Post* reported in 1990 that the former communist government of Czechoslovakia supplied at least **1,000 tons** of Semtex explosive to Libya. This may seem a small amount, but only approximately **200 grams** of Semtex was sufficient explosive to destroy **Pan Am Flight 103** in mid-air. By that calculus, there is currently enough Semtex to supply Libyan state-supported terrorists for the next 150 years. Some terrorist experts estimate that the true amount of Semtex available worldwide may be as high as **40,000** tons.

The **black powder pipe bomb** is the most widely used terrorist explosive device today. Its ease of manufacture (from a black powder base), low cost and wide availability all contribute to the prevalence of the explosive.

The following chart (Table III) shows effective radii for high explosive fragmentation bombs:

Table III: Bomb Damage Radius					
Weight of Explosive	Demolition Range	Irreparable Damage	Serious Damage	Minor Damage	Safe Distance
1-10 lbs.	3-5 ft.	5-9 ft.	20 ft.	100 ft.	900 ft.
10-25 lbs.	6-10 ft.	10-15 ft.	30 ft.	150 ft.	1,740 ft.
50 lbs.	12 ft.	23 ft.	50 ft.	340 ft.	2,140 ft.
150 lbs.	20 ft.	50 ft.	100 ft.	650 ft.	3,180 ft.
250 lbs.	30 ft.	60 ft.	120 ft.	800 ft.	3,720 ft.
500 lbs.	45 ft.	95 ft.	190 ft.	1,120 ft.	3,800 ft.
1,000 lbs.	75 ft.	150 ft.	300 ft.	1,600 ft.	3,800 ft.

Small Arms

Terrorists continue to use small arms (**pistols** and **handguns**) to carry out their operations. These weapons are common in **assassinations** and **hostage-taking incidents**, largely because terrorists can easily conceal them. Small arms are affordable, and ease of access makes them the weapon of choice for groups operating with relatively low financial and training ceilings.

Most pistols are **self-loading** or **semi-automatic**. However, some newer weapons, such as the **Tech 9**, are fully automatic. This advance in technology offers terrorists a high rate of fire and increased effectiveness.

Automatic Weapons

These are anti-personnel weapons, but they can also penetrate the skin of commercial aircraft and police vehicles. Most terrorist groups prefer such weapons because of wide availability, easy concealment and their high rate of fire. Furthermore, automatic weapons carry intense **psychological power** with the general public and law enforcement officials.

There are two types of automatic weapons currently available: the **assault rifle**, and the **machine-gun**. Gun dealers, shows and outright theft are all means to easily acquire automatic weapons. Currently the Soviet **7.62 mm AK-47 assault rifle** is easily the most popular automatic weapon among terrorists. It has a proven track record, and it is available in either **fully automatic** or commercially available **semi-automatic** versions. It is furthermore very simple for a terrorist to convert the semi-automatic version of the AK-47 into a fully automatic weapon.

Terrorists also often use **heavy machine guns**, especially the **0.50 calibre** and the Soviet **12.7-mm** heavy machine guns. Both are effective for assaults against lightly armored vehicles, aircraft and as sniper rifles.

Recent developments in technology have led to the production of smaller automatic weapons, which are becoming increasingly popular with terrorists.

Grenades

When they use grenades, terrorists usually employ **improvised grenades** in attacks, although there are cases in which terrorists have acquired common military grenades via various weapons networks. When terrorists have limited access to military grenades they often employ homemade devices that are of simple construction but quite effective. One such device consists of three sticks of **dynamite** or **plastic explosive** wrapped in nails. While crude, such devices are highly effective and deadly.

Weapons of Mass Destruction

The threat of terrorist acquisition of WMD presents a clear danger to the security of Western countries. Many technologies associated with WMD programs, particularly those associated with **chemical-biological weapons**, have dual-use civilian and military applications. Similar technology is useful both in the production of WMD and medicines, for example.

This paradox has led to increased technology proliferation, and increased the likelihood that terrorists and related groups may acquire the technologies to develop and manufacture these types of weapons, or the weapons themselves. Chemicals used to produce **nerve agents** also have applications used to make **plastics** and process **food products**. Modern pharmaceutical technology has dual-use capabilities to produce **biological warfare agents** as well as **vaccines** and **antibiotics**.

Chapter VIII: Entertainment Facilities

Introduction

Planning Steps

Step One: The Planning Process

Step Two: Planning Checklists

Step Three: Response Folders

Step Four: Initial Actions On-Scene

Immediate Actions

Bomb Threats and Bomb Incidents

Possible Weapon Types and Tactics

Entertainment Facilities

INTRODUCTION

The entertainment industry (television, motion pictures, theater, professional sports, the circus, festivals/carnivals, etc.) changes its operations from day-to-day to keep current with world activities and national situations. The likelihood of some sort of criminal or terrorist attack is therefore a real possibility. In order to become a **hard target** to terrorists, in comparison with a **soft target**, facility planners should establish a proactive response plan to meet these changes.

As with previous chapters, entertainment facility planners and management should establish a four-step planning process.

PLANNING STEPS

1. Begin a proactive planning methodology.
2. Build **response folders**, critical documents for emergency response.
3. Establish a structure for incident response.

Step One: The Planning Process

The goal of a "living" planning guide is to provide the planner with the information and necessary tools to plan, prepare, respond to, and then integrate planning with all public safety organizations that will play a role in the event of an attack on a facility.

The entertainment industry must consider the **objectives** of their attackers or terrorists. The objective may be to get the industry to **publicize** a terrorist group's achievements, in order to sow panic and fear.

■ *With this in mind, the industry requires a very proactive approach to the planning process — especially in relation to the dissemination of public information through the media.*

Management and planners need to understand, as they begin the planning process, that first responders who respond to a terrorist incident may harbor some initial **reluctance** as they arrive to respond to some forms of **media** (television news, for example).

The assembly of a proper plan, and effective coordination with relevant safety organizations will ensure a **proper response**, and the efficient utilization of resources during a developing terrorist incident.

STEP TWO: PLANNING CHECKLISTS

1. Developing an Emergency Operations Plan (EOP).

It will become clear in the early planning stages that most **Emergency Operations Plans (EOPs)** have a similar layout. Typical plans assign responsibility to an organization or to individuals, in order to carry out specific actions at projected times and places during emergencies. **An EOP sets organizational relationships and authority guidelines**. It also details how responders will protect people and property in a given situation. An EOP will describe how facility personnel will integrate and utilize equipment, personnel, resources and supplies during an event. It will also establish authorities for particular functions during an emergency. Finally a good EOP will address **mitigation strategies** following an event.

2. Conduct a Threat and Vulnerability Assessment of the Facility

A. Threat Assessment

The first step in the planning process is a determination of the level of the threat in the area. A comprehensive threat assessment will heavily impact the degree to which **management** exercises caution. A threat assessment will also determine the amount of protection that management will procure to deter **terrorists**.

Table I: Terrorist Threat Assessment	
Existence	A terrorist group is present in the area or region.
Capability	Terrorists have either demonstrated the ability to carry out an attack in the area, or another organization has assessed the capability of a terrorist group.
Intent	There is evidence of demonstrated terrorist activity, threats or actions in the area.
History	There has historically been demonstrated terrorist activity in the area.
Targeting	Current credible information exists on terrorist activities indicative of preparations for specific terrorist operations. This could include terrorist intelligence gathering, the preparation of devices, or other actions.
Security Information	Terrorists have gathered information about the internal politics and security considerations of a target, which impact on the capability of the terrorists to carry out their mission.

Determining the threat level in a given area is not a difficult process. **Open sources** (media, library, Internet, etc) of information exist and can aid in determining the threat level. Another source of information can be the development of an **interagency working group**. This group should include both public and private agencies concerned with terrorist threats in the area.

Follow certain guidelines while performing a threat assessment, outlined in Table I.

B. Conduct a Vulnerability Assessment

Develop a list of vulnerabilities from local resources or from the attached list (Table II). Seek the advice and assistance of the planning team to ensure the development of the most comprehensive list possible, to better prepare for an attack.

		Table II: Sample Vulnerability Assessment
❏	1.	Minimal security.
❏	2.	Easy access to the public.
❏	3.	Mixed business interaction with minimal contact. An example of this might be a clinic or office that operates in a location away from the secure hospital and has no reception point. This provides a terrorist the ability to conduct surveillance and gain intelligence without being challenged.
❏	4.	Central receiving. Personnel assigned to central receiving may be less likely to question suspicious packages because they are unfamiliar with the size and types of packages the organization receives.
❏	5.	Multilevel buildings. Facilities with many levels present a challenge to all planning documents. First, they are difficult to evacuate in any situation. Second, they provide an opportunity for a terrorist to hide devices or themselves. Third, elevator shafts and stairwells can carry contaminants through a building while individuals attempt to evacuate.

		Table II: (continued)
❏	6.	Groups in an area.
❏	7.	Parking garage.
❏	8.	Restricted exits/fire escapes.
❏	9.	Inadequate air handling systems.
❏	10.	No control of HVAC or other utilities in the building.
❏	11.	On site storage of HazMat and Bio-waste material.
❏	12.	Close proximity to other industrial businesses.
❏	13.	Routes of passage for hazardous material in close proximity. Does a major trucking route pass directly in front of the location?
❏	14.	Easy escape route via major roadway.
❏	15.	Concealment locations in close proximity. This could allow an attacker to hide his/her location in an attack and to execute a secondary attack.
❏	16.	No visible security presence.
❏	17.	Routine response to unscheduled events. Do personnel take events for granted? Do they respond to "bomb hoaxes" with reluctance?
❏	18.	Do personnel challenge unscheduled or unannounced visitors?
❏	19.	Has there already been an attack or incident of some kind?

		Table II: (continued)
❑	20.	Does the organization sponsor or support an event that may be controversial?
❑	21.	Is operational capacity beyond the abilities of security personnel?
❑	22.	Is the organization impenetrable?
❑	23.	Are personnel members learned in surveillance methods?
❑	24.	Is there a reporting method for the organization regarding unusual events?
❑	25.	Does a management team routinely inquire about strange occurrences at the organization's facilities or locations?
❑	26.	Is there intelligence coordination with other utility providers in the area?
❑	27.	Is a special event scheduled at one of the organization's facilities?
❑	28.	Is the organization covering an event, which external groups or individuals may scrutinize?
❑	29.	Have external parties sent bomb threats, anthrax hoaxes, or other hoaxes to the organization or to one of its facilities?
❑	30.	Does senior management participate in the planning process, both for special events and for day-to-day operations?

After the completion of a vulnerability list, write the questions in one column (as described in **Chapter IV: Hospitals**). The planning team should conduct an assessment using the vulnerability list. Place a check in the column applicable to the target. Total check marks in each of the columns. The higher the column total, the more susceptible is the target to a terrorist attack (Table III).

Table III: Sample Planning Checklist		
Location Name (eg News Room)	Location Name (eg Sports Arena)	
✓	❏	Minimal security.
❏	✓	Easy access to the location.
❏	❏	Central receiving.
✓	❏	Multilevel buildings.
❏	✓	Groups in an area.
❏	✓	Parking garage.
❏	✓	Restricted exits/fire escapes.
❏	✓	Inadequate air handling systems.
❏	✓	No control of HVAC or other utilities in the building.

		Table III: (continued)
❏	✓	On site storage of HazMat and Bio-waste material.
❏	❏	Close proximity to other industrial businesses.
✓	❏	Routes of passage for hazardous material in close proximity.
✓	❏	Easy escape route via major roadway.
✓	❏	Concealment locations in close proximity.
❏	✓	No visible security presence.
❏	✓	Routine response to unscheduled events.
❏	✓	Do personnel challenge unscheduled or unannounced visitors?
✓	❏	Has there been an attack or incident of some kind?
✓	❏	Does the organization sponsor or support event (s) that may be controversial?
7	10	**Total**

STEP THREE: RESPONSE FOLDERS

As explained in earlier chapters, a comprehensive **response folder** that covers all aspects of the facility can help to identify system weaknesses, and help to establish a template for first responders who respond to an attack at the facility.

RESPONSE FOLDER CONTENTS:

SECTION ONE

1. General site information:
    a. Role of the facility in the community.
    b. Economic impact on the local area.
    c. Employment.
2. Location:
    a. Detailed strip maps to the location from the nearest airports, interstate highways and other principal routes.
    b. Surrounding area or local area maps (counties, cities and towns within a 30-mile radius).
    c. Maps to additional facilities or support facilities.
3. Location boundaries:
    a. Area.
    b. General data.
4. Facility description:
    a. Photos.
    b. GPS coordinates.

    c. Map references.
    d. Azimuth and distances to local features.
    e. Grounds description.
    f. Helicopter landing zones with approach obstacles (trees, lights, radio towers and wires).
    g. Facility construction.
        i. Roof entrances.
        ii. Roof type and strength.
        iii. Fire escapes and ladders.
        iv. Stairways.
        v. Elevators.
        vi. Loading docks.

5. Security:
   Protective measures.
        i. General defensive strategies.
        ii. Protection strategy.
        iii. Security force in facility.
        iv. Security communications.
    b. Restricted areas.
        i. Reason for restriction.
        ii. Security measures in restricted areas.
    c. Security systems.
        i. Intrusion detection.
        ii. Surveillance measures.
        iii. Fire monitoring system.

6. Electrical power sources:
    a. Security system power supply.
    b. Backup power.
    c. Operations power.

   d. Diagrams (panel locations).
   e. Matrix.
       i.   Area/equipment affected.
       ii. Breaker.
       iii. Panel.
       iv. Panel locations.
   f. Impact of loss of electrical power.
7. Ventilation systems with photos:
   a. Air-conditioning unit.
   b. Size.
   c. Alarms.
   d. Monitoring.
8. Telephone system:
   a. Panel locations.
   b. Phone numbers to outside lines.
   c. Servicing company and point of contact.
   d. Cell phone availability and location.
9. Medical support:
   a. Location.
   b. Capabilities.
10. Fire protection:
   a. On-site firefighting equipment.
   b. Sprinkler system.
   c. Extinguisher locations.
   d. SCBA capabilities and locations.
   e. Hydrant locations.
   f. Halon capabilities.
   g. Mutual aid.
   h. First-in unit response times.
   i. Established command cost locations.

11. Hazards:
    a. Toxic chemicals.
    b. Biological waste.
    c. Infectious waste.
12. Water:
    a. Potable water location.
    b. Water shut-off locations.
    c. Emergency storage capabilities and storage locations.
    d. Maintenance company.
13. Sewer:
    a. Manhole locations.
    b. Routing diagrams and dimensions.
14. Vehicles:
    a. Types and number usually found.
    b. Locations of parking lots.
    c. Diagrams and capacities of enclosed or underground parking facilities.
15. Staffing:
    a. Key personnel data.

SECTION TWO

1. Building layout — floor diagrams:
    a. Doors.
    b. Interior windows.
    c. Rooms.
    d. CCTV locations.
    e. Other pertinent security information.

2. Operational considerations:
   a. Likely hostage-holding areas.
   b. Potential adversary holding areas.
      i. Fields of fire.
      ii. Concealment of additional weapons or devices.
3. Critical paths:
   a. Critical path diagrams (with photos).
   b. Critical path doors.
      i. List of doors along path with descriptions.
      ii. Door information matrix.
         1. Interior/exterior.
         2. Glass size (wire).
         3. Width.
         4. Height.
         5. Thickness.
         6. Lock type.
         7. Door opening direction.
         8. Material.
         9. Keys and key control.

In addition to the establishment of a response folder, planners should also consider the placement of an initial command post. For off-site facilities, establish **primary** and **secondary** locations for the **Incident Command Post (ICP)**. This methodology will give planners the ability to decrease a utility's profile and minimize the risk of a terrorist attack.

IMPORTANT NOTE

Obviously, a facility's response folder is a valuable document for a terrorist or criminal planning an attack. Therefore, planners **must** keep this document under strict control, and make it available **only** to those individuals who have a required security need to review or examine the material within the folder.

Step Four: Initial Actions On-Scene

In the event of a terrorist attack, utilize the monitoring and alarm systems either already in place or established for a special event at the facility. In a potential terrorist situation, first responders must practice measures to protect themselves. Only then can they provide assistance during the situation. The first step in this process is:

STOP, LOOK AND LISTEN
1. Resist rushing into the incident.
2. Approach the incident from upwind.
3. Stay clear of hazards.
4. Be mindful of confined spaces.

After planners establish an **EOP** and a **response folder**, they should determine priorities in the event of a terrorist incident. The following list is a starting point to help planners achieve some or all of the required objectives:

♦ Notify emergency response agencies.
♦ Notify key hospital planning staff.
♦ Activate internal emergency response personnel and resources.
♦ Secure and isolate the incident area.
♦ Account for the safety of the internal and external population (external refers to people or personnel immediately outside of the facility).
♦ Establish the incident command post.

♦ Assess the situation.
♦ Begin evacuation or sheltering in place, if required.
♦ Provide updates to people located throughout the facility (rumor control).
♦ Recommend and coordinate personnel protection measures.
♦ Provide overall coordination of the incident.
♦ Establish a system to handle public inquiries.
♦ Establish operational periods for staff, and relieve no essential staff, if able.

Zone Control

Regardless of whether the type of terrorist attack is relatively normal (**armed attack, bombing**), or an unusual occurrence (**WMD attack**), planners and responders must establish initial areas for continued safe operation.

Planners must consider both short-term response operations and extended operations that could last hours or days. While planning for zones of operation always establish:

1. Safe ingress to CP and staging locations.
2. Safe egress of emergency vehicles.
3. Control of ingress and egress around the facility staging areas.
4. Evacuation areas and routes.
5. Inner and outer perimeters.
6. Media holding areas.
7. Hot, Warm and Cold Zones for contamination.
8. Decontamination locations.

IMMEDIATE ACTIONS

When responders and planners arrive on-scene following a potential incident, the **correct identification** of the nature of the incident is the first step in first response. There may be an initial period of **chaos**. This will generally begin following an incident, when most people are in some form of **shock**. It is during this time that responders and planners must quickly take control.

Gain Control Of The Situation (GCOTS)
1. Size-up the situation.
2. Evaluate the situation.
3. Set incident priorities.
4. Determine the potential for further harm, injury or destruction.
5. Establish one incident command post.

During incident size-up, goals should be to provide answers to three questions:

♦ What is the present situation?
♦ What is the predicted behavior?
♦ How will this affect my incident priorities?

■ *If there is no life hazard, rescue situation or fire then there is no reason to risk exposure to emergency response personnel.*

	Table IV: Initial Response Action Checklist
❏	Park vehicle upwind and not too close to the scene.
❏	Establish a command post separated from direct involvement with casualties and personnel.
❏	Determine the size and magnitude of the affected area.
❏	Provide a situation report: 1. Number of casualties. 2. Location of inner and outer zone. 3. Possible incident scenario (bombing, WMD, armed attack).
❏	Request additional resources if needed: 1. More EMS/police/fire/HazMat/bomb squad. 2. Personnel recall if necessary. 3. Notification of hospitals.
❏	Establish radio control and a direct link to the Emergency Operations Center (EOC) if activated.
❏	Assign additional units to the command post.
❏	Determine if evacuation is necessary.
❏	Determine potential hazards to responders (secondary devices, chemicals, snipers).
❏	Request that a department command official respond to the location.
❏	Request that life services be disconnected, if necessary.
❏	If there is a chemical or smoke hazard, restrict helicopter access to the scene.
❏	Establish triage, treatment staging, sectors in the outer zone.

Table IV: (continued)	
❏	Request transportation for victims, by-standers, personnel.
❏	Establish decontamination, if necessary.
❏	Practice crime scene tactics and ensure security of the scene.
❏	Establish rehab areas and sectors for personnel.
❏	Establish a Public Information Officer (PIO).

Bomb Threats and Bomb Incidents

Terrorists can deliver bomb threats in a variety of ways. Most terrorists today **call threats** in to the target organization. Sometimes they may also communicate a threat **in writing**, or they may elect to play a **recording**.

There are two reasons to explain why a caller might report a bomb threat. **1)** The caller may be the person who placed the device, or the caller may have knowledge that a terrorist has or will place a bomb. He/she may want to minimize personal injury or property damage. **2)** Alternatively, the caller may want to create an **atmosphere of anxiety and panic** that will result in a **disruption** of the normal activities at a facility. Whatever the reason for the bomb threat, management must react to the possibility of a bomb attack. Proper planning can ensure that the variety of **uncontrollable reactions** to bomb threats is minimal.

Preparing and Planning for a Bomb Event (Actual or Hoax)

The establishment and employment of security measures and plans can help reduce the accessibility of a facility or building. Planners can identify those areas they can **harden** against potential terrorist attacks. If a bomb event occurs, prior planning will have instilled confidence in the leadership, by reinforcing the impression that those in charge care. Proper planning can also reduce the likelihood of panic. **During a bomb threat, panic is usually the primary**

achievement (and perhaps the goal) of the caller. As planners prepare to cope with a bomb incident, they should develop two separate but interdependent plans. These include a **physical security plan** and **a bomb incident plan**.

Bomb incident plans outline the procedures that management should implement when an organization receives a bomb threat, or in the event of an actual bombing. **When planning for a bomb event, establish a definite chain of command or line of authority**. A clearly defined line of authority will instill confidence and prevent panic.

It is easy to establish a chain of command if a facility has a simple office structure, and if one business occupies one whole building. However, if there are multiple building occupants, then representatives from each occupant entity should attend a planning seminar. Participants should **appoint a leader** and **delineate a clear line of succession**.

Designate a command center or other focal point for telephone and radio communications. Only permit individuals with assigned duties into the center. Establish provisions for alternates in the event someone is absent at the time that a terrorist calls in a threat. Keep an updated blueprint or floor plan of the facility in the command center.

(For further information about bomb threat planning and management response to bomb threats, as well as search teams, see **Chapter II: Security Planning**).

POSSIBLE WEAPON TYPES AND TACTICS

Letter and Parcel Bombs

The terrorist use of letter and parcel bombs usually heralds a specific campaign against a given target. Letter and parcel bombs serve two offensive functions. **First**, terrorists can design these weapons to kill and maim a target. **Second**, terrorists can use these bombs to harass the general public. The most common explosive used in the construction of letter and parcel bombs is **C4**. Terrorists can mold C4 into paper-thin sheets, which simplifies the challenge for the terrorist to conceal the bulkiness and weight of a device. **Micro-chip detonators** are currently available worldwide, and as a result, parcel or letter bombs today may have no distinguishing characteristics.

Another type of letter and parcel bomb in use today is the **Zip Gun**. This explosive device comprises three tubes that each contain 0.22 caliber bullets. The terrorist constructs a device that aims these bullets at the vital organs of the intended victim. The device triggers when an unsuspecting victim opens the package.

Vehicle Bombs

There are generally two types of vehicle bombs. The first is the **Ulster Car Bomb**. Terrorists who employ this tactic plant an explosive charge in, on or underneath a vehicle. Terrorists who employ this bomb type generally intend to carry out an **assassination**. Such bombs can explode when a victim opens a door or turns a key. The terrorist may also remotely detonate the device. This form of vehicle bomb was very popular with the **Provisional Irish Republican Army (PIRA)**.

The second type of vehicle bomb is the **Lebanese Car Bomb**. Terrorists typically construct this device inside a car, van or large panel truck. Terrorists who employ this bomb type can **1)** shape or construct a device within the skeleton and hull of an automobile, or **2)** build an extremely large device in the back of a van or truck. The use of vehicle bombs remains a popular terrorist tactic today.

Small Arms and Automatic Weapons

Terrorists also continue to use small arms, such as pistols and handguns, to carry out operations. They commonly use these small weapons for **assassinations** and **hostage-taking incidents**, largely because such weapons are easy to conceal. Small arms are relatively affordable and easy to procure.

Automatic weapons are anti-personnel weapons that can also pierce police vehicles and aircraft, if necessary. Terrorists groups gravitate towards automatic weapons because they are easy to acquire and conceal, and have high rates of fire.

Grenades

Terrorists employ both military and improvised grenades in their terrorist activities. When groups are unable to procure military grenades, they frequently build devices using primitive materials. Such devices may consist of taped sticks of **dynamite**, or **plastic explosives** wrapped in **nails**. These weapons, while quite crude, can be fatal and highly effective.

Chapter IX: Special Events

Introduction

Terrorism and Special Events

Public Order and Special Events

Sports Violence and Special Events

Planning for Special Events

Threat Assessments

Minimum Intelligence Requirements

Disorder and Special Events

Crowd Composition

Chemical, Biological and Radiological Attacks

CBRN Incident Indicators

Special Events

INTRODUCTION

Special events include any preplanned events that require security, law enforcement, fire, emergency medical or other operational resources **greater** than those needed to fulfill normal daily response requirements. Special events can include **music concerts**, **carnivals**, **circuses**, **sporting events**, **demonstrations**, **parades**, **races**, **strikes**, **conventions**, and political events such as **government meetings** or **summits**. Special events range from small events, which require minimal modifications in daily deployments, to large and complex events that force authorities to conduct **comprehensive pre-planning**. Large, complex events often require special, **dedicated planning staffs**, require resources from multiple agencies and disciplines, and may require staff augmentation through mutual aid or hiring of temporary staff. Examples of large, complex events include the **Olympic Games**, the **World Cup**, the **Super Bowl**, the **Stanley Cup Finals**, **International Naval Reviews**, and annual parades such as the **Rose Parade** or **St. Patrick's Day parades**. Events that have political or international significance can be among the most complex.

TERRORISM AND SPECIAL EVENTS

Historical experience highlights the importance of special events security planning. The **1972 Olympic Games** in Munich, West Germany saw the tragic massacre of 11 Israeli athletes. In this event, the **Black September** terrorist group seized control of the athletic dormitory in the Olympic Village, immediately killed two athletes and seized nine remaining hostages, and demanded the release of 200 Palestinian prisoners. Authorities refused to meet the group's demands, but gave the group safe passage to Furstenfeldbruck Airport. A gunfight ensued with police, and the remaining hostages, one policeman, and five terrorists died in the clash.

The **1996 Olympic Games** in Atlanta confirmed that the threat of terrorism at special events has not abated. That event involved the distribution of over 12 million spectator tickets at multiple venues. Although 75 per cent of the venues were located in a three-mile diameter 'Olympic Ring' in downtown Atlanta, the sole terrorist attack of the Games came in an area **outside** the main venues. The Centennial Park bombing, which occurred just past one o'clock in the morning on 27 July 1996, killed one person and prompted numerous bomb hoaxes in the aftermath of the bombing.

PUBLIC ORDER AND SPECIAL EVENTS

Political meetings are also frequently subject to public disturbances. Recent examples of these phenomena include the **World Trade Organization (WTO)** Ministerial Meetings in Seattle, Washington between 29 November and 3 December 1999. The meetings led to a series of week-long civil disturbances.

The WTO meetings involved 8,000 delegates from 135 nations, numerous **Non-Governmental Organizations (NGOs)**, and dignitaries that included US President Bill Clinton and US Secretary of State Madeleine Albright. Officials arrested 587 demonstrators for activities related to the disturbances.

Sports Violence and Special Events

Sporting events have historically led to significant mass violence and disturbances. In 1985, while hosting the **European Cup**, Heysel Stadium in Brussels witnessed a riot in which 39 people died and 437 sustained injuries. In 1990, Detroit was the scene of riots after the hometown Pistons won the **NBA Basketball Finals**. Revelers in Detroit rioted, leaving 8 people dead and 99 others wounded (26 from gunshots). There were 150 arrests. Furthermore, during the 1993 **Stanley Cup Finals** between the Montreal Canadiens and the Los Angeles Kings, a sports-related disturbance erupted in Montreal. After Montreal won the title, Canadiens fans rioted, looted 100 stores, and destroyed 15 buses (disrupting transit operations). The rioting led to 168 injuries and 115 arrests, and also caused US$10 million in property damage.

PLANNING FOR SPECIAL EVENTS

Each special event has its own unique needs, and triggers unique threats and law enforcement concerns. This is equally true for both **one-time** and **recurring** (e.g., annual) events. Weather and the level of popular interest also frequently influence attendance at special events. Criminal activities such as **pickpocketing**, **purse snatches**, **illegal vending**, **sales of counterfeit merchandise**, and **ticket scalping** are common features at special events.

■ *As a matter of course, special events attract large numbers of people who are usually willing to spend large sums of money.*

■ *This large attendance and level of media attention provide interest groups with a forum to express political dissent, or to garner publicity for a special interest.*

Accordingly, protests and demonstrations often accompany special events. The **political orientation** of the demonstrations need not necessarily be related to the event theme. When demonstrations are related to the event in question, however, authorities and planners should take heed. Some demonstrations have the potential to attract **counter-demonstrations**, particularly if protesters demonstrate on emotionally charged issues.

■ *Counter-demonstrations have the potential to escalate into confrontations and disturbances.*

Pre-planning can significantly improve the chances that law enforcement will successfully provide adequate security and public safety service at special events. Successful planning is dependent upon a recognition of certain key issues at special events. These issues include **coordination with events staff**, **traffic control**, **crowd control**, **medical assistance**, **personnel concerns**, **political considerations**, **business considerations**, **media relations**, and **comfort issues**.

Coordination with Events Staff. Special events have their own organizers and event staff to conduct event-related tasks. **Security** and **public safety personnel** must closely coordinate their activities with the event staff. Event staff typically control event operations, may have decision authority to conduct evacuations or ejections of attendees, and staff access control points.

Traffic Control. All special events have traffic issues, vehicular or pedestrian. **Adequate access and egress are requirements.** Parking areas and traffic flow must be pre-planned to avoid **congestion** and **gridlock**. Security of parking areas is necessary to minimize vehicle theft and burglary. Planners must provide emergency access for emergency vehicles, security forces and dignitaries. They must also pre-designate evacuation routes in case of emergencies, such as a fire or an attack.

Crowd Control. Crowd control is a pivotal issue in virtually all special events. The location or event venue, as well as the composition of the crowd, influence the authorities' range of crowd control options. Signs and the visibility of security or of the police presence are also important elements of crowd control. Planners must include contingencies for crowd dispersal, public order or disturbance management.

Medical Assistance. Medical assistance at large gatherings or special events is known as **mass gathering medicine**. Planners for special events must include both first aid and emergency/mass casualty options, and consider medical aid stations, specialty medical teams, pre-staged ambulances, and medical caches. Mass gathering medical personnel should be versed in major incident management, communication skills, and the safety aspects of stadium/venue design.

Mutual Aid. The need to address the myriad of contingencies that can arise during a special event, frequently exceeds the personnel and equipment capabilities of a single agency. Planners need to address provisions to access mutual aid resources prior to the start of the special event. These provisions should stipulate the **type** and **quantity** of **mutual aid**

available, provisions for activation and demobilization, identification of staging areas, types of missions to be assigned and accepted, and identification of **liaison personnel** and **command authority**.

Personnel Concerns. Special events usually require more personnel than are available through local resources. These personnel often work on overtime, while the sponsoring organizations typically reimburse expenses. Management must make provisions to track personnel expenses, as well as the use of vehicles and equipment.

During extended special events, management must **rotate personnel**, allow for **breaks**, and **feed personnel**. Organizers may issue special equipment to personnel and if so, planners should make provisions to account for and retrieve this special equipment once the event concludes.

Finally, **supervision is critical**. Staff may not have day-to-day experience in the special event environment. **Spans-of-control** must conform to the workload and the importance of assigned missions. Written post instructions may be necessary in complex settings.

Political Considerations. Special events may become a focal point for demonstrations or political action, and counter-demonstrations may also be a factor. Planners should consider steps to separate highly emotional or confrontational counter-groups. Identify key demonstration organizers in advance.

Business Considerations. Illegal vending, particularly sales of counterfeit event-related merchandise, will often be present at special events. The principal issue involved in this practice is **trademark infringement**. Vendors may need to procure a permit, and there may be prohibitions on the sales of certain items, such as **alcoholic beverages**, or **glass containers** that crowds could use as projectiles in a disturbance. Ticket scalping is another common practice. Provisions for screening concessionaires may be advisable.

Media Relations. In large events there will certainly be a large media presence. Planners should therefore be sure to develop a media liaison plan. Such plans should address logistical issues, such as the designation of areas for media vehicles where they will not obstruct access to exits or critical areas.

Comfort Issues. Access to toilets, seating and drinking water are important concerns at large scale special events. If possible, de-centralize comfort facilities, in order to **avoid crowding** and potential crowd surges, and to minimize the potential for pickpockets to congregate. Ease of access to water, particularly at outdoor events, can have important preventive benefits. Widespread availability of water can reduce incidents of **heat casualties**, which in turn have the potential to **strain medical resources** to limits.

THREAT ASSESSMENTS

The development of a proper special event organization requires a threat assessment. A special event threat assessment should cover both the scope of the event and likely external threats. The first consideration is the event's visibility. High visibility events require more resources. The assessment should list whether the event will take place at a single venue or if there will be multiple venues. Finally, the assessment should consider security risks. Typical security risks include:

♦ Protests/demonstrations.
♦ Arson/fire.
♦ Assaults.
♦ Building occupation.
♦ Terrorism (conventional terrorism, CBRN terrorism, cyber-terrorism, and hybrid forms of terrorism).

In the development of a special event plan, the agency has two primary goals: **1)** management of the special event, and **2)** prevention of a degradation of service in the primary event jurisdiction.

Special event planning should include a series of planning sessions with all involved entities, including the jurisdictional **law enforcement agencies (LEAs)**, fire service, events staff, in-house and external security agencies, participating groups, and adjacent jurisdictions. Key planning issues the sessions should address include:

Key Special Event Planning Issues
♦ Police/LEAs:
  • Crowd control, tactical response, on-site EOD/bomb squad, mounted policemen.
♦ Private Security:
  • Personnel, CCTV.
♦ Medical.
♦ Communications.
♦ Tickets/access/signs.
♦ Verification of credentials.
♦ Facilities/equipment/barricades:
  • Bull horns, shelters, traffic cones.
  • Lighting.
♦ Parking.
♦ Concessions/food service/water.
♦ Sanitation/portable toilets.
♦ Transport.

During the planning phase, it is essential that planners catalog the **agencies/organizations** that will be involved in the special event. Planners should pay particular attention to the responsibilities of various organizations, their capabilities (including limitations and restraints), resources, and command and control architecture (specifically command pathways).

The results of the planning sessions should be reflected in the event's **Emergency Operations Plan (EOP)**. Planners should identify venues, venue commanders, field forces and any specialized resources. Planners should conduct site surveys and develop

target folders/response information folders for each venue. Management and planners must also ensure there will be ample personnel for the event.

Planners should describe the existing security/public safety framework and address evacuation and dignitary protection issues. Finally, planners should also establish provisions to demobilize personnel once the event is complete.

Pre-planning should determine policies for **crowd dispersal**, **arrests**, and **use of force** (including crowd control agents and less-lethal agents or munitions) and clearly **articulate** and **communicate** these policies to all personnel involved in special event activities.

Planners must either obtain the equipment and materiel necessary to sustain operations, or establish mechanisms to rapidly procure such materiel should the need arise. Finally, planners should ensure that all personnel practice all special event measures in **exercises** prior to the start of the event.

Communication is an essential element of all special event operations. The EOP should detail all potential communications issues in advance, including the availability of a given radio frequency, usage and restrictions, and communications requirements and limitations.

Planners must make provisions for sufficient phone lines to accommodate surge capacity. This should include **long distance access**, **cell phones** (including dedicated cell sites), **satellite phones**, and **wire-**

less connections to the Internet. Planners should recognize that the media may tie up access to cell phone lines should a **newsworthy event** occur. Therefore, planners should reserve back-up capabilities to address this possibility.

Contingency planning should also address the impact that **inclement weather**, **extreme weather** or **threats of violence** may have on the event, and possible event **closure** or **cancellation**.

Access control issues include ticketing, verification of credentials, procedures to screen visitors, and screening areas. Pre-planning should address the likelihood of **resistance to screening procedures**. Establish additional procedures to address this possibility in-place. **Screening areas should maximize surveillance and minimize choke-points, which can become difficult to control**. Planning to address bomb threats, suspicious packages, and suspect devices is also necessary. For large scale events, consider pre-positioning **EOD/bomb squads** prior to the event.

MINIMUM INTELLIGENCE REQUIREMENTS

♦ The absolute minimum information necessary to plan for a special event includes:
♦ The location(s), time(s) and date(s) for the special event.
♦ Estimated attendance figures.
♦ The presence of special interest or security threat groups.
♦ The presence of large sums of money or valuable property.
♦ Previous event history.
♦ Information on the event organizers and the event staff.
♦ The presence of alcohol or glass containers.

Information about previous event history is especially vital. Information on past crowd composition and activity, including the presence of special interest groups, demonstrations and counter-demonstrations, can **help planners to craft response options**. Awareness of the presence of VIPs, dignitaries and protecting organizations is also crucial.

If there are protecting organizations present, information on their **protective details**, **escorts**, **motorcades**, **support staff**, and **protective measures** is essential. Examples of protective measures that protecting organizations may employ include underground or covert arrival/departure and departure times and locations, buffer zones, and concealment.

Planners should also identify the need for **airspace** and **waterway restrictions**. A checklist for special event intelligence preparation for operations/planning follows:

Table I: Intelligence Preparation for Operations: Special Event Planning
❏ Location (venues, terrain features/terrain analysis, route reconnaissance).
❏ Duration/time frame (day, night or 24-hour).
❏ Symbolic value (historical/political significance).
❏ Size/attendance (invitation only, free or paid admission).
❏ Media attention.
❏ Prior history (attendance, crowd composition and protests).
❏ Potential for disturbance (non-violent or violent demonstrations and counter-demonstrations).
❏ Crowd considerations (crowd composition and crowd profiling).
❏ Access control (vehicle access, stand-off areas/buffer zones, restricted airspace and waterways).

Using this information, the planning staff can assess the need for pre-positioned resources, including designated incident support sites (with alternates) such as command posts, staging areas, decon corridors, casualty collection points, treatment areas, evacuation

Chapter IX: Special Events

routes and assembly points. Detail and keep this information in venue-specific **target folders/response information folders**. Route reconnaissance (recon) is an important element of both **pre-event and operational intelligence** for special events.

■ *Special events and related activities will often alter typical traffic patterns.*

A template for conducting **route recon** follows.

Route Recon Template

Determine the location, and assess the impact of:

♦ Venue access/egress
♦ Security zones
♦ Street closures
♦ Scheduled/unscheduled protester activities
♦ Motorcades
♦ Altered travel patterns

Determine the travel time (under various conditions) to:

♦ Hospitals/trauma centers.
♦ Key support facilities.
♦ Airports, heliports, helispots and landing zones (LZs).
♦ Courts, jails and booking facilities.

DISORDER AND SPECIAL EVENTS

As mentioned earlier, special events often attract protesters and demonstrations, which may lead to disturbances. As special event planners develop crowd management options, they should consider environmental factors such as venue design. If possible, remove movable items, employ video (CCTV) monitors, and provide adequate signs and lighting in all areas. Consider factors such as highly visible security and police deployments.

Depending upon the nature of the threat, police should consider **mounted (equestrian)** and **mobile field force (platoon)** deployment options. In case of a disturbance, deployment configurations and tactics should be situation specific. Police may need to employ mobile tactics, dismounted tactics, and less-than-lethal options.

Active intelligence is essential to effective management of both crowds and disturbances. **Field observers** should gauge the situation, and provide field intelligence and real-time situation status reports to support tactical decision-making. **Should a disturbance occur, management must assess the situation, and contain, isolate, and disperse the disturbance**. Identification of outlets for crowd dispersal is a priority. In addition to the size and location of the crowd, essential elements of information at the **incident commander** level include:

♦ Proposed dispersal route.
♦ Current axis of crowd movement.
♦ Closed areas.
♦ Crowd composition.

CROWD COMPOSITION

The determination of the nature of the crowd, or *crowd composition*, is an essential element in the formulation of crowd control strategies and tactics.

There are eight basic crowd types. Four of these constitute **crowds**, while the remaining four constitute **mobs**. Mobs are more difficult to manage than crowds. The eight types are:

♦ Casual crowd.
♦ Cohesive crowd.
♦ Expressive crowd.
♦ Aggressive crowd.
♦ Aggressive mobs.
♦ Expressive mobs.
♦ Acquisitive mobs.
♦ Escape mobs.

Casual Crowds. This is the simplest grouping of people. Composed of individuals with no common interest or purpose, people within **casual crowds simply happen to be in the same place at the same time**. Casual crowds demonstrate an extremely low emotional level. Members within casual crowds **see themselves as individuals, not as members of a group**. It requires substantial provocation to stimulate violence within this type of crowd.

Cohesive crowds. These are assemblies of people who congregate together for a defined purpose. Members still consider themselves as individuals and not as members of a group, but **the crowd may possess intense internal discipline**. One example of a cohesive crowd might be spectators at a sporting event or concert. **Cohesive crowds often display high levels of emotional energy**. They infrequently erupt into violence.

Expressive crowds. This type of crowd possesses a unified expression of sentiment and frustration. A common purpose binds its members, who also seek leadership. **Members display a range from resignation to highly agitated levels of emotion**. Expressive crowds can quickly go into action if they become agitated. **An inability to communicate their dissatisfaction can lead to frustration**.

Aggressive crowds. These crowds have definite, strong feelings, which members express through a unity of purpose. **Individuals largely surrender their own identities to embrace the group's sentiment**. Participants are impulsive and emotional, and they are capable of following others into destructive and lawless behavior. **This is the most dangerous form of crowd since it can rapidly transform itself into an aggressive mob**.

Aggressive mobs. The aggressive mob engages in some form of violent or lawless behavior. This violence the mob may direct towards either persons or property, or both. **Riots often erupt as the means through which aggressive mobs release pent-up anger and emotion**. Emotion primarily motivates these mobs, and their actions are usually short-lived.

Expressive mobs. These mobs also seek to release pent-up emotions. **Members view violence as a legitimate means to publicize a cause or grievance**. Members often become frustrated and will demand a platform to air their views. **Expressive mobs can be unreasonable, and characteristically make outrageous demands**.

Acquisitive mobs. These are mobs that seek to acquire something, as in the case of looters who exploit the **chaos** and **confusion** that result from an existing riot. Since greed motivates these mobs, the resulting riots are often of a longer duration than other outbreaks. **Authorities usually find it easier to control acquisitive mobs than other mob-types, since they have little emotional commitment**.

Escape mobs. Panic often drives these groups, who might include people who flee from a fire or a collapsing structure. Since flight impulses motivate escape mobs, they can be especially dangerous. **This type of mob's activities can very quickly escalate**

beyond the control of police or security forces.
Escape mobs frequently cause stampedes. For example, an escape mob crushed 95 football fans in a
Hillsborough Stadium tragedy in the UK in 1989. A
1993 New Year's celebration in Hong Kong killed 20
people and injured 100 others, after a surge of 20,000
pushing and shoving revelers escalated into a
stampede.

CHEMICAL, BIOLOGICAL AND RADIOLOGICAL ATTACKS

Terrorists may employ chemical, biological or radiological weapons during a special event to cause mass casualties and mass disruption, or else to sow fear and terror. Special events may concentrate a large number of people in one place. Most large special events receive intense media coverage, ensuring that terrorists will have an audience for their attacks. **Additionally, special events may have political, social, economic or religious significance to the group that conducts the attack**. Identification and recognition of incidents involving these weapons is more complex than incidents that involve explosives. Each type of agent, whether chemical, biological or radiological, possesses unique characteristics.

Chemical attacks are more predictable than biological attacks in their results, and they generally yield an immediate impact. Biological events are more difficult to discern because an extended **incubation period** can mask the use of some agents. Many analysts believe medical response is likely to initially misdiagnose bio-attacks as suspicious outbreaks of disease.

Biological attacks take hours (in the case of toxins) to days or weeks to yield effect. This can make immediate recognition problematic, unless there is direct evidence (the discovery of dispersal devices), or unless management receives a threat communication.

■ *Persons infected during a biological attack are not likely to uncover the symptoms of their illness until after they leave the focal point of a special event.*

Nuclear attacks are the least likely form of terrorist attack. However, there will probably be no difficulty discerning the cause of such attacks. Most likely, a nuclear attack would take the form of a large blast.

Radiological attacks involve radiological dispersal devices (RDDs) and are principally tools of disruption. Radiation is invisible, odorless and tasteless. The discovery of a radiological attack would therefore be difficult, complicated by the fact that there is a **delay in the onset of radiological symptoms**. Unless responders employ radiological survey instruments (detectors), recognition of a radiological attack would probably depend on the discovery of direct evidence (e.g., an RDD or radio-luminescent material) or the receipt of a threat communication from a terrorist.

Planners should prepare checklists to ensure that **public safety**, **security** and **law enforcement personnel** who will respond to special events can recognize a chemical, biological or radiological incident. The following checklists summarize the indicators of chemical, biological and radiological incidents for response personnel and commanders:

CBRN Incident Indicators

Chemical Incident Indicators

Minutes to hours...

Unusual dead or dying animals
— Lack of insects
Unexplained casualties
— Multiple victims
— Serious illnesses
— Nausea, disorientation, difficulty breathing, convulsions
— Definite casualty patterns
Unusual Liquid, Spray or Vapor
— Droplets, oily film
— Unexplained odor
— Low-lying clouds/fog unrelated to weather
Suspicious Devices/Packages
— Unusual metal debris
— Abandoned spray devices
— Unexplained munitions

Biological Incident Indicators

Hours to days...

Unusual dead or dying animals
— Sick or dying animals, people or fish

Chapter IX: Special Events

Unusual Casualties
— Unusual illness for region/area
— Definite pattern inconsistent with natural disease
Unusual Liquid, Spray or Vapor
— Spraying and suspicious devices or packages
Unusual swarms of insects
— Vectors
Suspicious Outbreak of Disease

Radiological Incident Indicators

Delayed onset of Symptoms...

Unusual numbers of sick or dying people or animals
— Symptoms of radiation exposure
Unusual metal debris
— Unexplained devices/munitions-like material
Radiation Symbols
— Placards on container
Heat Emitting Material
— Emitting heat without signs of an external heating
  source
Glowing material/particles
— Radio-luminescence

Chapter X: Response
Response to a Terrorist Incident

First Response Incident Phases

First Response Incident Phases

Initial and Response Action

Management of a Terrorist Incident

History of Incident Management
and Command

Incident Commander Role and
Responsibilities

First Response Incident Phases

INTRODUCTION

After the commission of a terrorist incident, **first responders** will be acting on their own in the short term until additional assets arrive on-scene. With this in mind, responders should learn certain steps to facilitate incident management and achieve a minimum loss of life and property, as well as minimize first responder casualties.

Terrorist incidents, whether armed attacks, bombings or WMD incidents, also present challenges for **emergency management** officials. Post-incident response involves multi-agency operations, and this makes emergency **incident management** in terrorism situations unusually complex problems.

Emergency or disaster situations can overwhelm first responders and emergency management alike. This chapter is designed to give emergency personnel a guide to dealing with terrorist situations.

FIRST RESPONSE INCIDENT PHASES

Response to most emergency incidents occurs in four distinct phases (Table I).

Notification Phase
The notification phase begins when a dispatch center receives initial reports of an incident via a 911 call. The phase continues until the first units arrive. **First responders will usually not be aware that the incident to which they are responding is a terrorist attack beforehand**, barring prior intelligence. Initial notification usually arrives from individuals untrained in relaying important information to emergency operators. Such information includes the presence of mass casualties, multiple devices and/or chemicals.

Responders gather information in the notification phase, including the location of the incident, estimated numbers of victims, description of injuries as well as any incident indicators which would lead the call taker to suspect a terrorist attack.

Response Phase
The second phase during the incident is the most important for first responders. It begins with the initiation of incident management techniques: **scene assessment**, **safety** and **triage**.

The response phase of an incident is critical because it is then that emergency responders are most vulnerable to **secondary terrorist attacks**.

Table I: Incident Phases			
Notification Phase	**Response Phase**	**Recovery Phase**	**Restoration Phase**
Begins upon notification.	Begins with scene management.	Begins when all victims are removed.	Begins when the investigation is complete.
Lasts 15 to 30 minutes.		a. Take charge of Incident Command System. b. Scene security. c. Evacuation. d. Triage.	Designed to: a. Re-establish services. b. Activate state and federal services. c. Re-supply response crews if necessary.
Ends when scene management begins.	Ends when all casualties are removed; lasts two hours generally.	Can last generally 4 to 6 hours.	Can last for months if necessary.

Terrorists employ tactics designed to lure emergency responders onto a scene, and launch direct or time-delayed attacks while responders respond to the earlier incident. It is thus critical that responders become aware of the fact that the situation to which they are responding **may only be the first in a series of terrorist attacks** at the location.

Caution!

First responders should learn to avoid tunnel vision. For example: In a 'No-Notice' simulated sarin gas attack on the New York City subway system in April 1995, first responders arrived on-scene at a simulated Manhattan subway station explosion. Initial fire department units jumped to a 'no smoke, no fire' conclusion and proceeded into the station wearing minimal protection. As the firefighters entered the station, they encountered victims (actors) in varying states of distress. Exercise observers then informed the responders that they had exposed themselves to sarin agent, and were now casualties. The simulation death toll was 100 first responders.

Key points during the response phase include scene assessment, safety, commencement of rescue procedures, secondary device search, and institution of the **Incident Command System (ICS)**.

Recovery Phase

The recovery phase typically begins with the removal of the last casualty from the scene. Depending upon the size and scope of the incident, however, this phase may begin while some casualties remain on-scene.

The primary focus during the recovery phase is the restoration of essential services within the affected region or city sector. Recovery can include vehicle decontamination, boarding up windows, restoration of utilities and reinstatement of security in the area.

Restoration Phase

The last phase begins following rescue, investigation of the primary scene and reoccupation of locations and facilities. The primary emphasis during this phase is the restoration of the incident location to its original state. It is during this time that the healing process also begins within the affected community.

Initial and Response Actions

Responders and emergency call takers should follow certain procedures during the primary two phases of an incident.

911 Operators and Dispatchers

Emergency operators undergo extensive training in telephonic emergency medical instructions, equipment operation, and a host of interviewing techniques to elicit information from distressed individuals. Operators must be able to recognize telephonic descriptions of a possible attack, and know what questions to ask: *"Was there an explosion? How many victims are visible? Is the caller seemingly affected in any way? What is presently occurring?"*

During the **initial notification phase**, frantic calls for help will overload the emergency operations center, and responders must anticipate this eventuality. For their part, operators and dispatchers must gather and disseminate relevant information to ensure that first responders receive as much information as possible about the scene of an incident prior to arrival.

To ensure that emergency response crews receive the best possible information relevant to a situation, **standard operating procedures (SOPs)** should keep emergency operators and dispatchers in the loop regarding intelligence information about potential or threatened terrorist attacks. These are the individuals who will most likely receive calls after the commission of a terrorist attack.

The primary functions of the emergency operator and dispatcher are to:

♦ Elicit relevant information regarding the incident from the initial caller.
♦ Provide pertinent information relevant to safe approach and scene status (chemicals).
♦ Remain calm and know how to activate his or her agency's Anti-/ Counter-Terrorism Plan (ACTP).

Responder Initial Notification Actions
Emergency responders should assess the nature and description of the call, and if necessary alert additional units and wait for instructions. Responders should locate the best possible access route to the scene, starting **at least two blocks** from the incident. If the area is unfamiliar, the responder should seek assistance.

Wind is an important consideration. Regardless of the incident description, responders should suspect an airborne hazard and should approach the scene from upwind. They should consider the equipment needed to protect response crews (turnout gear, Self-Contained Breathing Apparatus (SCBA), level A Haz-Mat suit).

Responders should anticipate the possibility of multiple hazards, natural and man-made. There could be a **secondary explosive** or a **chemical** or **biological device** present on scene. Likewise, terrorists can be among the injured or lurking close to the scene. First responders should practice four initial response actions:

1. Evaluate and take control of the scene.
2. Secure the scene.
3. Evacuate the area.
4. Begin triage.

REMEMBER:
First responders are potential secondary targets.

On-scene, the first arriving units should immediately establish scene control techniques to ensure public safety. Responders should evaluate the size of the incident area, the severity of damage and the number of injured people. Remaining emergency responders and the overall **incident commander** should then begin to isolate and take control of the situation. Controlling the scene, isolating hazards, triage and evacuation are all resource-intensive. Therefore, first responders should **immediately** determine the amount of help they require on-scene.

To better coordinate the response effort, first responders should consider establishing **outer** and **inner perimeter zones**. This can help to develop a sense of the size of the situation and the operational capabilities of the units on scene.

Establishing Perimeters and Control

It may be difficult for units to establish perimeter control measures due to the nature of the attack (large bomb, WMD), as well as panic among bystanders and victims attempting to get away from the scene.

By adequately sizing up the scene and evaluating potential hazards, the initial units can establish 'stand-off' or 'work zones' around the scene. Depending upon the size and magnitude of the situation, responders may wish to divide the zones into an **inner zone** and an **outer zone**.

The Inner Zone

The inner zone isolates the incident and any potential associated hazards. Within the inner zone, responders can carry out specific rescue operations and also observe ongoing operations. **One or two access control points** can restrict movement into the inner zone. An example of the kind of hazard that might dictate an inner zone is an explosion, or the discovery of a possible **secondary device**.

The Outer Zone

This safe area is the farthest control access point or boundary of the incident area. Responders limit and control all access to the scene through this location. The 'outer' zone provides a safe location from which responders and incident commanders can effectively plan, deploy, treat and respond to victims.

■ *An initially wide outer zone, which responders later elect to shrink, is easier to execute than an initially small zone that responders later need to expand.*

The initial outer zone in the Oklahoma City bombing, for example, was 20 blocks from the building. The following are considerations for inner and outer control zones.

The amount and type of resources on hand provide an estimate of the practical size and associated actions for each zone.

♦ Capability of available resources will also affect the zones. Responders should restrict access into the inner zone, and keep out untrained people.
♦ No matter how well-trained, if individuals cannot protect themselves (with HazMat, helmets, masks) they should not enter an affected zone.
♦ The size of the incident also shapes the size and scope of the zones.

After responders assess the initial incident size and magnitude and establish working zones, they should take the following response steps to secure the scene.

Securing the Scene

As with any type of emergency incident, scene security assumes a vital role. Law enforcement and possibly other forms of access control resources (private security, National Guard) can establish a **security perimeter**. This will eliminate the possibility of:

♦ Destruction of property.
♦ Destruction of evidence.
♦ Scene convergence by unwanted personnel.
♦ Unauthorized access by unauthorized personnel.
♦ Looting.

During the initial response, responders may identify certain facilities and locations that require further protection. The IC may relocate security resources without depleting the overall incident of protection, after he/she establishes a security perimeter around the incident.

Should I Evacuate the Area?

Evacuation may be necessary to protect the public at large. The issue of evacuation is generally a difficult one for most responders. Frequently responders and the IC on-scene want to protect the public by removing them from a hazardous area. However, possible **exposure** stemming from the removal of a citizen or victim from an affected area **can cause further harm**. Such hazards can include nuclear, chemical, biological, explosive and/or incendiary devices.

If first responders feel evacuation is necessary, they can execute one of three evacuation procedures: a total evacuation of all threatened populations; in-place sheltering for all areas within or around the incident area; or a combination of evacuation and in-place sheltering within the incident area.

Evacuation

The selected option depends on what responders find and what responders fear. First responders initially count the number of victims, citizens and bystanders in the affected area. In certain types of attacks it may not be possible to determine the size of the affected area, given a lack of tell-tale **incident indicators** (smoke, flames and debris). In this situation, responders need to **estimate** the number of possible casualties and victims, and plan accordingly.

Once responders determine just how large a radius around the incident area they will evacuate, they must consider resources and weigh any extenuating circumstances that may be factors in an evacuation. Responders should remember that their primary response effort is the **immediate** incident at hand, while they are determining the resources needed. Most secondary units will still be attempting to reach the scene and assist victims.

■ *If responders intend to fully evacuate the area, they must state this intention early in the incident so they can allocate other resources to this phase of the response.*

Responders should query **local responders**, when determining the evacuation zone size, regarding locations such as nursing homes, hospitals or child care centers in the evacuation area. After they gather information as to the demographics of the area, responders must notify the people within the zone.

Responders will generally face three types of people in the evacuation zone. The **first** is the victim/ bystander trying to get away from the scene. **Second** will be the bystander/victim trying to get closer to the scene. **Third** will be a person in his or her own home/ apartment/office, either afraid to come outside or unaware that there is a potential danger in the area.

The first response agency's **ACTP** should enumerate the **notification** methods (local media, door-to-door, police public address system) used to inform the public. In the event of an immediate evacuation, a means to **transport** the evacuees is necessary. The agency ACTP should catalog available transportation resources (public and school buses, police vehicles) slated to evacuate the scene of an incident quickly, and stipulate a secure evacuation route.

In-Place Sheltering
Responders may need to employ in-place sheltering
— the restriction of citizens to their homes for safety
— if the situation presents an immediate risk to citizens or if there is no time to evacuate the area. The
difficulty with in-place sheltering techniques is they
generally require some training in proper techniques
prior to an incident.

Insufficient manpower and equipment can compromise responders' ability to establish in-place sheltering via the media. Instead, responders can convey
the following in-place sheltering techniques to citizens via their car radio PA system:

♦ Block any openings in rooms with tape or other
  material.
♦ Close all windows and ventilation systems.
♦ Shut off air intake systems, air conditioners and
  fans.
♦ Listen to the radio or TV for further instructions.
♦ Do not leave the house until told to do so.

**Combination of Evacuation with In-Place
Sheltering**
Both evacuation and in-place sheltering may be
appropriate during an incident. Such a situation might
occur when responding resources cannot support a
total area evacuation. If this is the case, responders
can systematically perform evacuations of the areas

at greatest risk, and provide in-place sheltering training or personnel support to other areas.

Use of this hybrid method constitutes a judgement call, an option during the **initial phase** of response. The most important considerations for any evacuation are quick decision-making and the presence of logistical obstacles, such as transportation and facility-specific impediments.

Triage

Triage is an initial response action. Regardless of the agency or department dispatched initially to the scene of a terrorist attack, the primary goal of all responders should be **treatment** and **assistance** to victims.

In order to ensure that all victims receive proper care, responders need to perform rapid **triage**. This process of rapidly classifying victims on the basis of their injuries will increase their chances of survival and allow responders to better manage their assistance.

Finally, despite the experience and level of training, first responders will almost certainly overlook an important element during a terrorist incident. Therefore it is advisable to establish three or four response actions, to establish control on-scene until the arrival of additional resources. **It is important to remember that first responders are not alone during an incident.** The checklist provided in Table II is designed to help first responders react usefully when they encounter an overwhelming situation.

Management of a Terrorist Incident

Effective management of a terrorist incident requires that the Incident Commander (IC) ensures a response capability, based on the day-to-day operations of those responding to the scene. This practical and systematic approach to incident management reduces confusion, inefficiency and the possibility of harm.

■ *The overall goal of incident management is to avoid making the situation worse than it already is.*

Table II: Response Action Checklist
❏ Park vehicles upwind and not too close to the scene.
❏ Establish a command post away from direct involvement with casualties and personnel.
❏ Determine the size and magnitude of the affected area.
❏ Provide a situation report: 1. Number of casualties.   2. Location of inner and outer zones.   3. Possible scenario of the incident (bombing, WMD, armed attack).
❏ Establish radio control and direct link to the Emergency Operations Center (EOC), if activated.
❏ Assign additional units to the command post.

Table II: (continued)

❏ Determine if evacuation is necessary.

❏ Determine potential hazards to responders (secondary devices, chemicals, snipers).

❏ Request that department command officials respond to the location.

❏ Request that life services be disconnected, if necessary.

❏ If there is a chemical or smoke hazard, restrict helicopters from the scene.

❏ Establish triage, treatment staging and sectors in the outer zone.

❏ Request transportation for victims, by-standers and personnel.

❏ Establish decontamination, if necessary.

❏ Establish rehabilitation areas and sectors for personnel.

❏ Establish a Public Information Officer (PIO).

Chapter X: Response

HISTORY OF INCIDENT MANAGEMENT AND COMMAND

The Incident Management System/Incident Command System (IMS/ICS) came into existence in the early 1970s to manage rapidly moving wildfires. At the time, emergency managers faced a number of problems, including:

♦ Different emergency response organizational structures existed.
♦ Lack of command and control.
♦ Incompatible communications equipment.
♦ Unclear mission objectives.
♦ Lack of a coordinated plan.
♦ Assignment of responders and equipment to many different supervisors and locations.

An interagency task force, part of a cooperative local, state and federal effort called FIRESCOPE (Firefighting Resources of California Organized for Potential Emergencies), developed the original ICS.

FIRESCOPE focused on four elements early in the process. **First**, the proposed system had to be flexible to meet the needs for incidents of all types or sizes. **Second**, agencies needed to be able to adapt the system on a day-to-day basis for routine situations or large-scale emergencies. **Third**, it had to become the standard, enabling responders from many different backgrounds to adopt the common structure. **Finally**, the ICS had to be cost-effective and practical.

As noted, the need for a response system for large-scale wildfires prompted the development of the ICS. However, from its development the ICS betrayed common characteristics with other emergency situations, including:

♦ Rapid incident development.
♦ Incidents occurred without notice.
♦ Incidents could potentially grow in size and complexity, if unchecked.
♦ Incidents presented high risk to personnel.
♦ Multiple agencies were on-scene, and each had some level of responsibility.
♦ Capacity to become a multi-jurisdictional event.
♦ High media visibility.
♦ Risk to life and property.

Local, state and federal agencies in the US today widely use the ICS. This standard management model allows responders to deal safely and effectively with the complexities of incidents of all sizes.

■ *No single agency or department can address and handle a terrorist attack alone. Various agencies, both public and private, must work together. This team approach helps to ensure the use of all available resources to their maximum benefit.*

The IMS/ICS Concept

The IMS/ICS structure is designed from a top-down perspective. Command functions are established with the arrival of the first unit, whose leader assumes the role of IC and starts to organize a response. Additional responsibilities in other areas thereafter may also fall under the IC. There are five functional roles of the IMS/ICS organizational structure (Table III).

Table III: Typical Organizational Structure	
Role	**Responsibility**
Command	Overall direction and control of the incident. May include management staff positions responsible for public information.
Operations	All initial response operations at the incident.
Planning	The collection, evaluation, dissemination and use of information about the incident development and status of resources.
Logistics	Supply of facilities, services and materials for the incident.
Finance/ Administration	All cost and financial considerations of the incident.

This recommended modular approach could change depending upon incident magnitude or operational necessity. **Remember:** base a command structure on incident criteria and needs, and not a prefabricated arrangement.

Incident Action Plan (IAP)
Each situation calls for an Incident Action Plan (IAP). This can be written or verbal and is recommended for incidents involving multiple agencies and jurisdictions, or if an incident is complex, such as the April 1995 Oklahoma City bombing, which killed 168 people. The IC or members of multi-agency organizations should develop an IAP to establish the priorities, goals and support activities of the operation.

Manageable Span of Control
A manageable span of control is the number of responders one supervisor can effectively manage. The recommended supervisor to personnel ratio is between 1:3 and 1:7, while an optimal ratio is 1:5.

Emergency Coordination Centers
During typical emergency responses, many agencies establish command and control positions of their own. Confusion ensues because each agency sees its own command post as the center of operations. Personnel thus become uncertain as to where they should report.

■ *After initial response to a terrorist attack, the designated IC should clearly determine locations (CP, Staging and EOC) that meet the needs of the incident, not the desires of individual agencies.*

Once this stage is complete, all coordination, direction, control and resource management should be directed from this central location.

Incident Response Process Using the IMS/ICS

The initial unit arriving on scene should **(1)** assess the scene and establish command and control measures, **(2)** establish scene security, **(3)** make decisions about evacuation and **(4)** begin triage.

After additional resources arrive, the incident management process can swing into further action. Management begins with an action process to establish control, and this requires the commander to immediately determine incident size and scope. This **stabilization** process follows the initial response phase.

INCIDENT COMMANDER ROLE AND RESPONSIBILITIES

In brief, the Incident Commander is in charge. The IC appointment generally rests with whoever has overall control of an incident, usually a local police, fire or emergency manager.

■ *In most cases, the first arriving police officer or fire company assumes the role of IC and will relinquish those duties upon the arrival of command staff members.*

In the United States, the Federal Bureau of Investigation (FBI) and the Federal Emergency Management Agency (FEMA) have responsibility as Lead Federal Agencies for crisis and consequence management for terrorist incidents in the US. **This, however, does not give either agency the assignment as IC.**

The IC is responsible for overall incident management, including:

1. Performance of command activities.
2. Protection of life and property.
3. Control of resources, including personnel and equipment.
4. Responsibility for the safety of response personnel and their mission.

5. The creation of an effective liaison between outside agencies and available resources.

Management Responsibilities of the IC

An effective IC must be proactive, decisive, objective, calm and quick-thinking. To handle all the responsibilities of this role, the IC also needs to be adaptable, flexible and realistic about his or her limitations.

The IC must be a **leader, not a micro-manager**. Typically, individuals prefer to perform an act themselves rather than delegate tasks.

Responsibilities of the Incident Commander:
1. Assess incident priorities.
2. Determine goals (longer term).
3. Determine objectives (shorter term).
4. Develop and implement incident action plans.
5. Develop organizational structure.
6. Manage incident resources.
7. Coordinate overall emergency activities.
8. Ensure responder safety.
9. Coordinate activities of outside agencies.
10. Authorize the release of information to the media.

1) Assess Incident Priorities

There are three major incident priorities: **life safety**, **incident stabilization** and **property conservation**. The first priority is always the safety of emergency response personnel and other citizens. Next is incident stabilization. The IC must develop a strategy to minimize the incident's effect on the surrounding area.

■ *The nature of the command system that the IC develops should be in keeping with the complexity of the incident, not its size.*

Even when situations appear hopeless the IC must manage and control them as far as possible. The final priority is **property conservation**: minimize damage and yet achieve pre-set goals and objectives.

2-3) Determine Goals and Determine Objectives
The IC must ensure the deployment of resources for maximum control. An IC must identify goals and translate them into objectives. **Goals** are in the overall plan used to control an incident. They are broad in nature and are achievable by the completion of certain objectives. **Objectives** are the specific operations that responders must accomplish to achieve their goals. Objectives must be both **specific** and **measurable**; one goal can have multiple objectives.

4) Develop and Implement the Incident Action Plan
The IC is the primary individual responsible for development and coordination of an incident action plan. The IC must ensure that the action plan addresses the needs and requirements of all agencies that respond to a terrorist incident.

5) Organizational Structure

As the complexity of an incident grows, the IC should implement an organizational structure from the top down, as needed. In this way, the commander can delegate functional responsibilities as they arise.

> ■ *Key to organizational structure is a manageable span of control. Effective management is difficult if too many people report to one supervisor.*

The IC should delegate functional areas to others within the system, decreasing the number of individuals that report to him or her directly. The IC must anticipate and prepare for span of control problems, especially during the rapid build-up of resources and personnel at the onset of an incident.

6) Manage Incident Resources

The initial determination of needs is only the first step in resource management. The IC must continually evaluate and adjust the deployment of resources, based on changes in the situation, contingencies, goals and objectives. As soon as the IC determines goals and objectives, he/she should evaluate the necessary resources and identify what is needed.

7) Coordinate Overall Emergency Activities

Coordination of overall emergency activities is vital to incident control. The IC can eliminate time wasted on unnecessary tasks if he/she can ensure the wise and effective use of resources. This requires continuous incident monitoring to ensure effective coordination, and to ensure that personnel do not duplicate effort or work at cross-purposes.

8) Ensure Responder Safety
Effective incident management requires a high personnel safety priority. Every responder necessarily serves as his or her own safety officer, yet the ultimate responsibility for on-scene safety rests with the IC. If there is an escalation or other contingency, the commander may direct a **Safety Officer** to ensure that responders are working safely and effectively.

9) Coordinate Activities of Outside Agencies
The IC is also responsible for the overall effective conduct of the response effort. Effective response requires multiple public and private agencies, and it is the IC's responsibility to maintain open lines of communication and coordination during a terrorist incident.

10) Authorize the Release of Information to the Media
The IC will play a major role in media relations, with responsibility for interviews and status reports during incident response. Coordination rests with the Joint Information Center (JIC), which is in turn under IC control.

The need for an effective IC cannot be overly stressed, particularly during a response to an unpredictable incident that can easily escalate out of control.

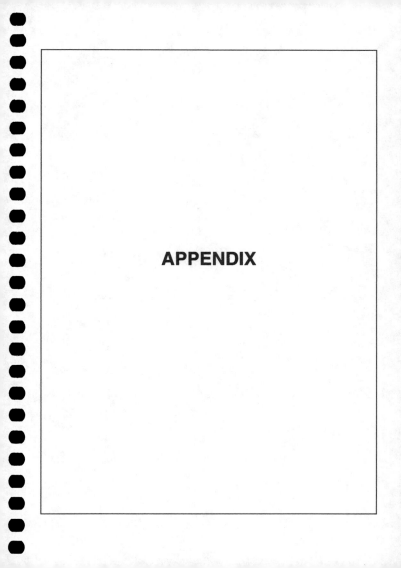

APPENDIX

Appendix A: Glossary of Acronyms

AC: Alternating Current
ACTP: Anti– / Counter-Terrorism Plan
ANO: Abu Nidal Organization
AP: Action Plan
ATF: Bureau of Alcohol, Tobacco and Firearms (US)
BR: Red Brigades (Brigate Rosse)

C4: Composition 4, or plastic explosive
CB: Chemical-Biological
CBRN: Chemical, Biological, Radiological or Nuclear
CCTV: Closed-Circuit TeleVision
COA: Course of Action
CP: Command Post
CPR: CardioPulmonary Resuscitation

decon: Decontamination
DHHS: Department of Health and Human Services (US)
DoD: Department of Defense (US)
DoE: Department of Energy (US)
DoJ: Department of Justice (US)
DoS: Department of State (US)
DoT: Department of Transportation (US)
DVD: Digital Video Disc

ELN: National Liberation Army (Ejército de Liberación Nacional)

EMS: Emergency Medical Services
EOC: Emergency Operations Center
EOD: Explosive Ordnance Disposal
EOP: Emergency Operations Plan
EPA: Environmental Protection Agency (US)
ETA: Basque Fatherland and Liberty [Euzkadi Ta Askatasuna (ETA)]

FBI: Federal Bureau of Investigation
FEMA: Federal Emergency Management Agency
FIRESCOPE: Firefighting Resources of California Organized for Potential Emergencies

GCOTS: Gain Control Of The Situation

HazMat: Hazardous Materials
HCF: Health Care Facility
HEPA: High Efficiency Particulate Arrestance
HVAC: Heating, Ventilation and Air-Conditioning

IA: Immediate Action
IAP: Incident Action Plan
IC: Incident Commander
ICP: Incident Command Post
ICS: Incident Command System
IMS: Incident Management System
IPO: Intelligence Preparation for Operations
IRA: Irish Republican Army

JIC: Joint Information Center

LAW: Light Anti-armor Weapon
LAW: Light Anti-tank Weapon
LEA: Law Enforcement Agency
LFA: Lead Federal Agency
LNG: Liquid Natural Gas
LPG: Liquid Petroleum Gas
LRA: Lord's Resistance Army
LZ: Landing Zones

MCI: Mass Casualty Incident

NBC: Nuclear, Biological and Chemical
NGO: Non-Governmental Organization

OEM: Office of Emergency Management

PAX: Passenger
PDD: Presidential Decision Directive
PETN: PentaeryThriol-TetraNitrate; also known as detacord, primex, or primeacord
PIO: Public Information Officer
PIRA: Provisional Irish Republican Army
PLF: Palestinian Liberation Front
POC: Point of Contact
PPE: Personal Protective Equipment

RDD: Radiological Dispersal Device

SCBA: Self-Contained Breathing Apparatus
SL: Shining Path (Sendero Luminoso)
SOP: Standard Operating Procedure

TNT: TriNitroToluene

VAV: Variable Air Volume

WMD: Weapons of Mass Destruction

Appendix